THE AUDACITY OF FUTURISM

THE AUDACITY OF FUTURISM

A PRAGMATIC APPROACH TO ENVISIONING THE FUTURE

R.S. Amblee

GLOTURE BOOKS

Grateful acknowledgment is made to the following for permission to reprint excerpts of their work: International Monetary Fund, Dow Jones and Company, Inc., World Health Organization, National Center for Public Policy and Higher Education, www.House.gov, www.cbo.gov, www.eia.doe. gov, www.nasa.gov, www.ers.usda.gov, www.oecd.org, and www.cms.gov.

ISBN-13: 9780983157427
ISBN-10: 0983157421
GLOTURE BOOKS

TABLE OF CONTENTS

ACKNOWLEDGEMENTS

There are several groups of people to whom I am very thankful for their help and advice. First, many thanks to Rusty Fischer whose valuable advice and organization of the chapters highlighted the principles on which the entire book is based. Many thanks to Cheryl Cook and Elandee Thompson for their feedback on the flow of the concepts. Thanks to Laurence Singer, David Hale, Kathleen Tracy, Alan Beck, Andrew Robinson, and Tom Kerr for their valuable suggestions and evaluation of the book. And thanks to my family and friends for their helpful comments, advice, and support.

INTRODUCTION

Looking into the future does not come naturally for most of us. In fact, it is hard enough for us to see past today. This book makes an effort to present a unique technique for looking into the future with real world examples. As we begin to visualize the future, we will see the human technological evolution is unstoppable and will explore how future generations must be prepared to face it.

During our discussion, we will not make unrealistic assumptions the world will experience some extraordinary technological breakthroughs, but rather will use whatever technology we have today—and more toil and sweat—at the research table to discover the inevitability of human progress.

Though our ancestors built highways and mega bridges with an expectation of a better future, they could see only a near future and not a distant one. The energy crisis is one such example. With industries built around oil and coal, previous generations possibly knew that one day these resources would not be able to support the huge global demand.

However, they never imagined underdeveloped countries would progress so fast. They were simply not able to estimate the size of today's global economy from the vantage point of those long-forgotten *good old days.*

The one guideline I use extensively in this book, and will continue to use in future work, is to look at our history not just for facts but also for a pattern.

This search for a predictable paradigm in human history is highly intuitive and provokes thoughts of futuristic possibilities. It allows us to develop forward-thinking capabilities and possible ways to predict the future of technological evolution.

Part I

The Fundamentals

CHAPTER 1

THE JOURNEY SO FAR

Ever since the proverbial sky began falling on poor Chicken Little, we have been hearing nearly endless scientific theories of how the world will end. Now, more than ever, most of those theories feature the devastating forces of black holes, gamma rays, asteroid strikes, global warming, and a nonstop list of other extinction-level threats straight out of last summer's million-dollar blockbuster films.

Unfortunately for our species, the list of dangers to our survival keeps growing as we find new technologies that discover hidden threats to humanity. From all outward indications, human beings are so fragile we do not stand a chance against such global disasters. Considering the grave nature of those considerable forces working against us, it is extremely fortunate we have survived for so long without becoming extinct.

While intellectuals who understand these very real, very scientific, and very provable threats feel understandably helpless, the public seems to pay little attention to the daily indicators of our impending doom. To them, it seems such stories more closely resemble Grandma's bedtime fairytales of old—a little fantasy, a little folly—quickly forgotten in the night's other restless thoughts. To the public eye, those threats attract the same attention as any other daily news.

Mass Extinction?

Protecting ourselves from any of these disasters is so unfeasible it would take billions of dollars to even investigate and understand the nature of the issues we face, let alone blueprint and implement them. Even though our technology appears to be advancing at a dizzying speed, when it comes to finding authentic solutions to these credible threats, we still have a long way to go.

Our governments cannot initiate any type of decisive action until there is sufficient proof to convince a majority of legislators to use public money to achieve such means. Private businesses, which are under no such restrictions, shy away from entering into the fray because the investments are still too risky to consider seriously.

So, what is to be our fate as individuals, as communities, as a species, as a planet? Will we all become helpless victims of mass extinction?

There is only one answer to this question: it all depends on where we are along our technological evolutionary path when we encounter those threats. Ideally, we will be at the right level of technological evolution to survive certain extinction-level threats. If we encounter them before we have sufficiently evolved to withstand them, we may not stand a chance.

On the other hand, if those threats come at a later stage, *after* we cross that evolutional demarcation point, we will not only survive, but also alter those situations to our advantage and indeed flourish in the coming age(s) of our existence.

Solar flares, as an example, are a threat to our ecosystem, but evolution creates strengths out of weaknesses. Imagine the ramifications if we could harness the sun's tremendous amount of energy. Technological evolution will make it so.

Volcanoes, forest fires, hurricanes, and tornados are all examples of huge energy sources we are unable to capture because we are simply not there yet along our evolutionary timeline.

We Have a Natural Desire to Evolve Technologically

Unlike animals that are forced to adapt physically to changing environments for survival, as human beings we bring that change upon ourselves by discovering new technologies through intellectual ability and adaptation to a higher standard of living. Rather than experiencing the hormonal fight-or-flight response of wild animals, we experience spy-it-and-buy-it instead.

Humans seem to have an almost insatiable desire to acquire *things* to lead a safer and easier, often upscale life. Even extremely wealthy people, who have all the creature comforts in the world, look for innovative items in the market to make their lives even better. The wealthy and elite are usually the first ones to buy new hi-tech, and high priced, gadgets when they arrive on the market. What is behind this craze for better living? Why are we not happy with what we have?

The answer to this question is in our natural desire to evolve technologically. We contribute to this natural evolution on a daily basis. Suppose you are in a wine shop looking for a particular vintage. Say you stumble upon a new design of bottle opener in your search for the perfect Pinot Noir. You wish to own it and get excited about buying it, even though you already have several bottle openers at home, all of which are more than sufficient for the job. The only things that may stop you from making the purchase are the quality and price. If the price is within your budget and the quality is acceptable, you will go for it. Every day we try to make life better than yesterday.

This natural desire toward the next big thing, the next level of technology, and the latest tools to make life easier is what brought us out of Stone Age caves and moved this evolutionary process to where it stands today.

Consumerism, Not Necessity, Is the Mother of Invention

Consumerism is what makes inventions successful. If a great innovation comes to the market but no one is interested in buying it, it is worthless. We give credit to great inventors for immeasurably improving our lives, but forget the end-users who created the demand for the product in the first place. For example, Thomas Edison is widely remembered as the inventor of the electric light bulb. What if no one wanted to use electricity and we were all just as happy with candlesticks and fireplaces? Without end-users to clamor for his brainstorm, like so many before or since, that invention would have vanished into thin air, along with Edison's estimable legacy.

Instead, the light bulb became a big success because of the common person's desire to use that technology to achieve a higher standard of living. Candles, matches, flames, and candleholders evolved to light bulbs, fixtures, light switches, and remote controls. As long as this desire to improve our lives continually exists, *human technological evolution simply cannot be stopped.* Inventors' innovative ideas are absorbed by the commercial market as quickly as they appear, although not all inventions are successful.

Even though intellectual efforts are important, the truth is, even if those scientists were never born, someone else would have created those inventions later. Human evolutionary activity is a collective effort and does not bear adherence to any one individual. The technological evolution is as natural as human life. Regardless of whether we consciously consider it, we are all part of this natural process.

Slow and Steady Wins the (Human) Race

If, on the other hand, an innovation were too futuristic for current times, it would remain on the back burner until its day finally arrived. Evolutionary changes have to be slow and steady to match the absorption capacity of the market; it takes time for people to adapt to these changes. For example, some of the inventions of Leonardo da Vinci never saw the light of the day during his own lifetime or for centuries after. They remained on paper waiting, in effect, for history to catch up with the great inventor. Now, after so many centuries, scientists are able to analyze and understand what he was trying to bring to life.

The same argument goes for many of Einstein's theories, from the theory of relativity to dark matter. Some theories are even too futuristic for us to understand today. As human knowledge advances, however, those futuristic concepts will be verified and eventually absorbed into the system. Until then, sadly, they will remain on the evolutionary back burner until we as a species catch up to them.

CHAPTER 2

UNDERSTANDING CONSUMERISM

H uman technological evolution is a process of discovering new technologies to satisfy human desires. It is not just the invention but the consumption as well. Throughout this book, the discussion focuses on *technological* not *biological* evolution. Henceforth, use of the term *evolution* refers to technological evolution.

All the present hi-tech marvels we enjoy today—from smart phones to laptops to whisper-quiet washing machines—exist because of the many geniuses throughout modern history that devoted their lives to inventions. Humans have responded positively by encouraging such inventions, either by buying them or investing in them.

Not all inventions become successful. Only some become very popular and even become mega industries employing thousands of people.

New ideas fail, if the product or service is not competitive in either price or quality or if they are simply not inspiring enough for people to buy them. This is the reason why thousands of new inventions vanish before they hit the market. The market absorbs only those technologies that are affordable and improve consumers' standard of living. Surprisingly, the failure of certain inventions does not dampen the evolutionary spirit. Instead, new companies and businesses keep popping up and the competition for survival continues.

Depending on what section of the human population a particular invention targets, the price and quality must be right for it to succeed. Obviously, this process of natural selection leaves many new inventions out in the cold.

Comparing the number of utility products in our homes today to what people had 100 years ago is staggering. By encouraging these products, we have come a long way in creating a web of industries and an enormous global economy that sustains, comforts, feeds, and enriches us on a daily basis.

Consumers always look for less expensive but quality products, and throughout commercial history businesses have always competed with each other to produce them. This partnership between producer and consumer is the backbone of human technological evolution. We evolved to our present state on this model and we will continue evolving on it into the future.

The technological evolution has not come without problems. Employees are at a higher stress level than just a few decades ago and many work several hours beyond their forty-hour-a-week limits to meet the crushing deadlines that are now a daily reality in order to produce products faster and cheaper.

Employees not only work harder every day but also have to increase productivity to reduce the company's overhead; multitasking to increase efficiency on behalf of their employer is now a job within a job. Every day employees have to prove they are still the best and make the company competitive in the market. If not, they will be replaced quickly for the simple reason that companies cannot afford to keep less skilled workers on the payroll when others will do the job more quickly or efficiently. After all, if they do, they will lose out to the competition.

If we examine it from an evolutionary standpoint, this complicated employee-employer partnership bears a close resemblance to the survival-of-the-fittest theory. Those who adapt, will survive; those who do not, will perish.

Why do we make our lives so stressful? Again, the answer to this question is in our natural desire to evolve. Humans have an intense desire to evolve so we are all equally responsible for such hardship. As an illustration, every time you shop for a less expensive product in a shopping mall or major department store, you are unknowingly responsible for creating competition in the marketplace. The very act of choosing a less expensive product over a more expensive one could potentially bankrupt a less competitive business or cause the lay-off of a less skilled employee. Many of us are unable to correlate the two, though there is a distinct relationship between them to even the most casual of observers.

Every time we go to a supermarket and pick a less expensive item of equal or superior quality as opposed to a more expensive one, we are indirectly responsible for putting stress on the employees that produce the more expensive item. Thus, the cheaper we buy, the more competitive those companies become. There is immense pressure on employees, as well as employers, to survive in this competitive economy, despite the fact we are all proud to live in such a civilized world.

This human technological evolution is unstoppable and its consequences are inevitable.

Discouraging Competition, for Better or Worse

Many countries have tried a different route to stem the tide of this blistering competitive marketplace and its effect on their economies. They have made regulations that discourage outside competition. They have insulated their local industries by not allowing outside products. Despite their best efforts, however, this fiscal protection for the local enterprisers and a stringent sense of job security for the employees over a long period has resulted in their economy lagging behind the rest of the world.

Lack of competition gives no motivation for businesses to grow. It will be inevitable that those governments will open their doors to world competition, so not only their local businesses will capture the world market but also invite global investments. It is a well-known fact an open economy is the fastest way to grow and the only way to survive in the real world. Every business is competing with the rest of the world; there is no time out for the competitive retailer.

Today, of course, many businesses are becoming global on day one. Internet e-commerce has made this happen with a technological evolution that has been both steep and unforgiving; seemingly, overnight the industrial world entered unblinkingly into the Information Age. Any successful invention quickly becomes a global product, thanks to the Internet; adapt or perish is once again the new battle cry for companies wishing to compete and, naturally, survive.

CHAPTER 3

THE NEED TO LOOK AT THE FUTURE

To endure in this competitive market, industries always look for three main resources -raw materials, energy and labor at as low cost as possibly feasible, to lower their cost of production.

When inexpensive raw materials are not available they look for alternatives. For example, if wood and steel become prohibitively expensive because of a sudden shortage, industries will use alternative materials like plastic, rubber, and other metal alloys. Having already gone through the plastic era in our lifetime, we have witnessed how plastic rapidly replaced many traditional raw materials. Though it is not a permanent solution, this Band-Aid quickly became an easy fix. Industries continually look for and research alternative materials to either make a product cheaper or better quality to win in the competitive marketplace.

Similarly, if labor becomes expensive in one area of the world, industries hire labor from other, less developed regions. This is called outsourcing, which takes employment away from one state or country and while rewarding those jobs somewhere else, often a world away.

Outsourcing has helped employees in less developed countries improve their standard of living. Ironically, many countries that used to provide low-cost labor are now becoming more expensive, which will prompt industries in developed countries to hunt for lower cost labor elsewhere.

The most important factor among all is energy, which affects industries globally. When the economy grows at a steady pace, the increased demand for energy makes it expensive and, in turn, increases the cost of all goods and services. When that happens, the buying capacity of people decreases. This slows down the economy. This fluctuation in energy demand/supply is one of the major causes of economic boom and recession.

Below is the world's energy consumption since 1775.

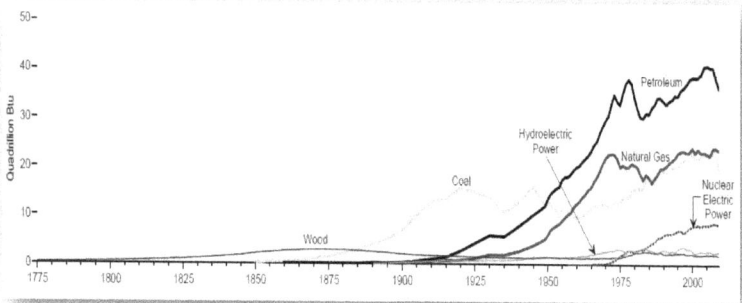

Fig. 1: Primary energy consumption by source 1775-2009

Humans depend on resources like coal and oil, which have huge production backlogs. For example, an oil refinery takes nearly a decade to complete a turnkey project from start to finish. With this kind of serious bottleneck in place, energy production always lags behind demand and quickly reaches the limit when demand exceeds supply during sudden, or even steady, economic growth. The graph below shows the possible effect of a hypothetical energy bottleneck on the global economy in the near future.

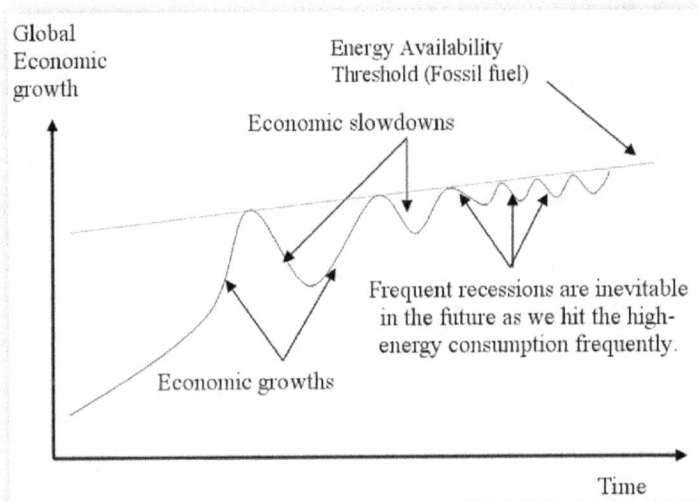

Fig. 2: Future energy bottleneck

With this energy bottleneck, our technological evolution is like a bird learning to fly. Until its wing muscles become stronger, it keeps falling. Unless we find a great source to produce energy, we will keep hitting this bottleneck and endure economic recessions.

In addition to energy bottlenecks, there are many other factors that are responsible for recessions, which can take years to recover from and more than a decade to regain lost economic momentum. The economy in the US, Europe, and other countries around the world has collapsed several times this century; this has been a painful experience for all of us.

Is This Economic Growth Worth It?

With the ebb and flow of our global economy, people are concerned about the future of the world as we know it. Looking at all the stress people go through during economic booms and suffering during recessions, it is natural for people to wonder, "Is this economic growth really worth it?"

The reality is there is no end to human desire; it only increases as our standard of living rises. A person with one television desires another; eventually those two televisions are not big enough, and so they replace them with two big-screens. Later, they may upgrade to high-definition. A person having one cell phone naturally wishes to have one for each family member. These are reasonable desires, considering our modern lifestyles. After all, as technology races ahead of demand, the products become increasingly affordable.

These desires are natural and inevitable as families, incomes, and expectations grow. As technology brings advanced products to market, our desire to own them naturally increases. In fact, we work even harder to acquire them. With an economy built on satisfying these desires, we have very little control over its growth.

Another interesting aspect of technological evolution is that growth is not linear. When it is slow, we believe the world is going to be like this forever, which places stress and frustration on us all. Despite working hard, it can take generations for families to rise out of poverty. They do not believe they are evolving, though they are, in fact, part of the evolution.

Looking at the bigger picture of human evolutionary history, the facts are amazing. From the cave man's way of life to a modern luxury home is a colossal technological evolution that took a very long time. Now, we are slowly beginning to understand this pace is not arbitrary or subjective; the pace is in our own hands. We can move at an amazing speed or pause it for a long time.

Though the human mind is extraordinarily intelligent, we often misunderstand each other. Egos clash. Emotions get the best of us, sapping our rationale. We let greed take over. Throughout our evolutionary history humans have feuded, envied, and betrayed. These shortcomings have led to isolationism, territorialism, and, occasionally, war and destruction. Technological evolution may move at a slow pace during difficult times—events like war and recession can slow it down or pause it for a while—but it is an unstoppable process.

As long as we are weak, the chances of surviving global disasters are bleak. Every day is important in our progress towards a strong future.

Since industrialization, the pace of technological evolution has increased dramatically. Our lives have changed noticeably, too. The changes the world faced in the past decade are significantly larger than any time in human history, with an exponential evolutionary pace.

If humans continue to evolve, what will our future look like? How will our standard of living be? Will we be more capable of handling a planetary crisis like global warming or a meteor strike?

Though these thought-provoking questions seem impossible to answer now, by understanding the forces behind technological evolution, we can potentially visualize how humans can—and will—progress to meet the inevitable challenges of the future.

Predicting the future relies on utilizing all available past and present data as a basis to develop reasonable expectations for the times ahead. Nowadays, large corporations invest heavily in future predictions to help direct resources towards possible business opportunities, risks, major economic events, and technological developments. Futurists are hired as personnel in many multinational companies.

Futures studies, a highly promising academic discipline by itself, focuses on pattern-based understanding of the past and present to forecast the likelihood of future events or trends. You may have noticed the plural term "futures"; this pluralization indicates the field's capability for envisioning alternative and even preferred futures.

There is a debate as to whether this discipline is an art or a science. In fact it is both. A rational future prediction involves knowledge and sound reasoning. Futures studies seek to understand what is likely to continue and what could plausibly change. In this highly competitive world such knowledge becomes essential to the survival and longevity of many companies.

This book examines the five major factors that primarily influence future trends. It makes an effort to establish strategies to predict the likelihood of future events or trends using these five primary drivers:

The Influence of Innovation
The Influence of Globalization
The Influence of Automation
The Influence of Technological Evolution
The Influence of Economic Recession

As we discuss and analyze these five influencing factors, we will more clearly see the path to the future, and feel the strength of evolutionary forces.

PART II
THE FIVE INFLUENCING FACTORS

CHAPTER 4

THE INFLUENCE OF INNOVATION

Understanding the Driving Force

Every time we go shopping, we notice a host of new products that are all products of intense human desire. The success of these products depends on how affordable they are and how much desire they satisfy for the consumer, be it a new smart phone or a video game or a tote bag. Human desire is the driving force behind all innovations leading to technological evolution.

To understand this reasoning, we need to focus on three basic criteria that show why products and services are invented and their future evolution. They are: invention/evolution, quality, and cost.

Invention/Evolution Criteria (I/E criteria)

The invention criterion is the reason why a product or service came into existence in the first place. It identifies the basic human desire it satisfies. The evolution criterion, on the other hand, shows how this product or service evolves into the future, satisfying additional human desires. For example, the invention criterion of the pen was to produce a written document for preservation. One of its evolution criterion was the ability to write continuously without dipping for ink; in other words to facilitate continuous writing. Other evolution criteria include a better grip, smoother writing, color ink options, and so on. The invention criteria and the evolution criteria together make up the utility value of the product or service.

A product or service will have only one invention criterion. However, it may have several evolution criteria like different grades, sizes, lengths, and so on to satisfy more human desires. The invention criterion of an automobile is to transport and its evolution criteria include interior comfort, music system, and greater speed. Early cars were very crude and only satisfied the invention criterion to transport. They were not comfortable, and certainly did not have anti-lock brakes or air bags. These primitive cars evolved gradually into modern cars to satisfy growing human desires. Remember, the invention criterion never changed in these cars, if it does, it will become another product.

When a product or service establishes in the market, it will undergo gradual evolution to satisfy more human desires. For example, if we desire a faster car, then *speed* becomes the evolutionary criteria and brings sports cars to the market, driving the technology behind a high-speed engine. Similarly, if we desire good music in the car, then an audio system becomes the evolutionary criteria that will drive the radio, CD player, and speaker products to evolve. If for some reason the majority of people had not wanted to listen to music while driving, then these automotive music products would not have appeared in the market. Now, after being in the market, each one of these products will undergo several upgrades as technology improves.

Another good example is a coffee maker. Its invention criterion was to satisfy our desire for a cup of freshly brewed coffee. As modern life became busier its time saving features influenced its future evolution. We wanted to make good coffee, but quickly. In older models, we had to heat water and pour it into the filter. Now, the machine heats the water itself, saving time.

Next, coffee makers attached to a water supply, eliminating the requirement to pour water, saving more time. Expect time saving features to dominate future modifications. Though there are many aesthetically designed coffee makers, if either of these two criteria—fresh brewed coffee or time saving—is compromised, they will fail in the market.

The next time you see a new coffee maker, notice its design criteria. If it is just shape, color, or some other irrelevant feature like an embedded digital clock or an alarm setting, and if it is slow in making coffee, it will not be in the market for a long time. It all comes down to how quickly the coffee is brewed and how little human effort is required.

What is the future of coffee makers? Already there are programmable designs allowing preparation of the coffee in the morning before you wake. In the future, there will be coffee makers with embedded computer chips, which understand our habits and prepare the coffee in advance with no programming needed. They will collect data on how often we make coffee and at what times, for a week or two, and then start mimicking our usage. Coffee makers will continue to evolve saving us more time and better understanding our habits to satisfy our desires.

The future of any home device, not just coffee makers, is moving toward becoming interactive, with an understanding of our usage patterns, not just to save us time but being proactive in their service.

Just like products, even services would evolve technologically. A good example is a restaurant. Traditionally, the invention criterion of a restaurant has always been to serve food. However, as economic activity increased in cities, an evolution criterion of *food availability* began to influence restaurants. This evolution criterion led to the mushrooming of restaurants, which now fill cities.

Further, as time progressed another evolution criterion, *time to meal*, began to influence the technological evolution of restaurants. This gave rise to two types of eateries: full service restaurants for people who want to spend more time—for business discussions, family gatherings, or holiday parties—and fast food restaurants for people like workers or travelers who are on the move and do not have much time to spend.

Let us see how these two types of restaurants evolved. In full service restaurants interior ambiance became an evolution criterion. Our desire for a cozy environment has brought in various types of interior designs and continues to evolve. For fast food restaurants, speed is of the essence. People want to pick up food and leave instead of dining in.

Many restaurants offer drive-thru windows to facilitate this time criterion.

Now let us use the strategy of driving force to look into the future. There will be a clustering of full service restaurants, similar to mini-food courts, giving customers more choices in one place to satisfy the food availability criterion. This food-cluster will resemble a regular restaurant, but one person will take orders for multiple cuisines. These food clusters answer the current lack of cuisine options. During family gatherings, each of us has our own preference in cuisine, but we all end up in one restaurant that caters only to some of us because the availability of multiple cuisines in any one place is lacking.

Looking further into the future, to address the interior ambiance criterion, clusters will use digital architecture rather than fixed decorations, where the décor theme of the restaurant can change each day, lending to a dynamic environment. Still farther into the future, customers will get to choose their own theme in individualized areas, such as choosing an ocean view or a mountain view. Even dining tables will become digital, offering world information right at the table, enabling you to even work while eating.

In the case of fast food restaurants—which have already adopted the clustering model discussed above for full service restaurants—more focus will be put on instant service to save us time. Whether ordering food from our cell phone or local kiosk, the wait time will be next to zero. Instant service will become the future evolution criterion. To make this instant service a reality, fast-food restaurants will be fully automated.

Quality Range Criteria and Cost Range Criteria

The above examples clearly indicate inventions and their technological evolutions follow human desires. Though new inventions are an attempt to increase the living standard of consumers, their continuous technological evolution requires an enormous amount of research, which needs investment so businesses follow the ladder principle to increase the quality of products and services little by little to suit the end users' spending capacity. As human desires are never ending, technological evolution is perpetual.

The extent to which the quality of a product or service will evolve depends on how much we are willing to pay for it. Let us say a store is selling a box of wooden toothpicks for 25 cents and other toothpicks made of high-quality material for five dollars. Most people will choose the cheaper one, as higher quality in toothpicks is unnecessary as long as they are hygienic. Even if the second company has invested a lot of money in improving the quality of those toothpicks—their texture, appearance, material used, and so on—it will still fail in the market or the market size will be very small, as quality and the cost are too high. If Apple releases its iphone models once every five years with multitude of features and with a hefty price tag it will fail in the market.

The question of how much quality we need in a product or service is where the quality-ladder comes into play. Each step of the ladder represents the quality-cost range, which is the buying capacity of the market. If there is a high quality but very expensive product that is well outside the range for the target market, that product or service will fail.

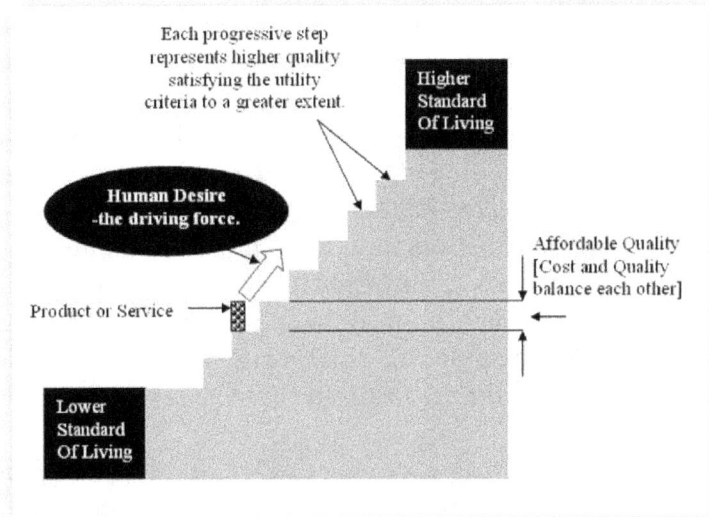

Fig. 3: Quality ladder

If you look at the above diagram, you can see the quality of a product or service increases a step at a time to suit the buying capacity of the market. The quality and cost ultimately decide the fate of the product. The quality must be acceptable and the cost must be affordable to the targeted end user.

Every time a new version of a product releases into the market, it will have a little more quality than the previous one to suit the buying capacity of the market. This is the reason we see digital equipment—camera, cell phone, laptop, video games, consoles—that is better year after year, in steady, small increments. Many times cost will never change but quality and features will. Companies are reluctant to invest too much money in a product, making them more expensive when they have the potential to fail. The point being, quality and cost balance each other out, and the product evolves in increments.

Now let's use this quality-ladder concept in the evolution of restaurants. Be it dine-in or fast food, our desire for higher quality and lower cost has brought about innovative cook-wear and appliances. Compare this to utensils used hundred years ago. If you look at the history of any restaurant, you will see a pattern of quality improvement. If readers could spend some of their spare time and do little more research, if would greatly help envision the future technologies that we could expect.

Quality of cooking equipment will continue to improve following the ladder principle, all leading to full automation.

Identifying the Driving Force
Sometimes it is hard to identify the driving criteria, be it the invention or the evolution criteria. However, if you look at the past history of any product or service, you will see a pattern of evolution. The focus of this book is to show the art of tracing the evolutionary path of products and services so we can grasp the forces driving this evolving world.

CHAPTER 5

THE INFLUENCE OF GLOBALIZATION

Understanding Globalization

Although it is market demand that decides the fate of the inventions, market size determines how fast they will evolve. Larger-scale global markets bring more revenue to invest in research for further improvement in technology, rather than smaller, local markets.

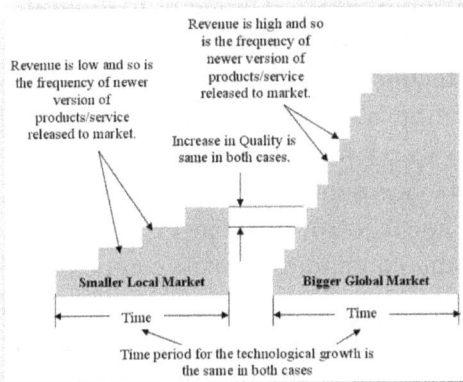

Fig. 4: Effect of market size on evolutionary pace

Globalizing a product dramatically increases the pace of its technological evolution. We can easily see this impact in the world of digital gadgets. The demand for better laptops and better cell phones is so high that billions of dollars are spent on improving them. This is possible because of the huge revenues earned from these products in the global market. Because the revenue is high, the competition is too. Every generation of digital devices is better than its predecessor. Continuous improvement is essential to survive in this competitive world.

This is the reason why consumers always look for brand products. They think larger, deeper pocketed companies do a better job in quality research and management. Just to remind you that research and development is a very time consuming and expensive process with no guarantee of final useful result. Breaking into well-established markets is a big hurdle for new companies. They have no other option but to look for new, innovative products that are not in the market. We have seen many instances of garage inventions developing into global businesses. The quest to innovate keeps engineers and entrepreneurs motivated to create new products, moving our technological evolution to higher levels.

Globalization is not simply selling products globally. It is a continuous search for better technology, better raw materials, better management skills (white collar jobs), lower cost labor (blue collar jobs), and bigger markets. Every one of these helps the companies remain competitive. A company may design a product in one country, manufacture it in another, and sell it in an entirely different country.

For example, many luxury cruise ships are designed in countries where design skill is available and assembled in

THE AUDACITY OF FUTURISM

another country where low cost labor is available. Finally, it may be operated as a cruise line in the Caribbean where the market demand is high. Regardless of the complexity of such an arrangement, it is most efficient for the cruise line company to do business this way.

There is another approach called *adaptive globalization* where countries export products that can adapt naturally to the target market so they become as effective as local products. For example, if a United States-made pizza is globalized and sold in Asia, it has to adapt to the local market to succeed. The obvious change is local flavor. When it adapts to the local market, suddenly it tastes very different. Likewise, a voice-activated GPS should understand the local language or exported movies might need to be dubbed to capture the interest of the local audience. To succeed, businesses have to adapt to target markets around the world.

Companies always compete to capture a bigger market share. To do that the cost has to be competitive. Typically, lower costs are achieved by mass production, which can be done in two ways: using low-cost labor or automation. As long as low-cost labor is available around the world, outsourcing happens naturally to bring down the product or service cost. Automation would follow suit.

Opening the Floodgate of Globalization

Globalization is not something new to the world. Trades have taken place throughout recorded human history. However, ancient trading was very small as it was merely an exchange of food, natural resources, and little handmade goods. There were no industrial products to trade. Increased international business activity occurred in the Industrial Age when products began to dominate the market.

While the daily consumption of food for a family does not vary much unless the family size itself increases, a similar constraint on the number of products consumed does not exist. Businesses that produce products also create jobs. Jobs bring more income to families. With higher income, people buy even more goods and the economy grows.

During the era of industrialization, even though there were plenty of products to be traded, the pace of trade depended on communication. If we look at history, as communication technology improved so did global trade.

Before computers, information flowed through communication media like telephones, newspapers, and radio as well as through business relations and personal contacts. Trading internationally was not an easy venture for new businesses.

After information technology (IT) evolved and computers came into existence, communication became much easier and brought globalization to a new level. The newfound information media affected the life of ordinary people, too. We now can buy or sell anything online. Many people conduct business from home with the help of computers. Imagine selling a clock or an art painting a couple of decades ago to a customer in another country yourself. Today this is possible because of IT.

If you look at the growth pattern of globalization, we notice it was actually the lack of information that was holding us back all those years from doing business with the rest of the world. Today there are billions of websites all over the world making the flow of information possible and finally opening the floodgate of globalization.

Land of Opportunities

Why certain regions of the world are favored as land of opportunities? Although human desire is the force behind economic growth, how much people are willing to spend to increase their standard of living, determines the size of the market and thereby the pace of economic growth. Though all humans have desires, people in developed countries tend to spend more.

The very reason developed countries evolved so fast is because when people are willing to spend more to improve their standard of living, a strong market emerges, which gives rise to businesses, creates jobs, spurs competition, and increases product quality.

People in developed countries spend more than people of underdeveloped ones but not because they have more. It is not the dollar amount that is important, but the ratio of what people earn to what they spend. If you look at some countries, people save so much that their financial conservatism slows their national economic development. Although the banks are flooded with cash from savings, the businesses suffer because consumers are spending conservatively. *Economy* by definition is the flow of money and not just the lump of cash. If money is stagnant, it is not productive. Similarly, *wealth* by definition is not just the resource, it is the demand for it. Demand comes from economic activity. Strong economy leads to increased wealth.

If an average family in one country spends a little more than another country, the economy in the first country grows a little faster than the second and over several decades this adds up significantly. It all comes down to how much people are willing spend to improve their standard of living from what they earn. This spending habit decides the pace of economic growth. A nation's economic status is in the hands of its own people.

Any country with a motivated market is a land of opportunity. Most developed countries are. New ideas and new inventions are highly appreciated in a motivated market. When people buy more, industries produce more.

The Fastest Way to Global Economic Growth

Although developed countries enjoy a higher standard of living, they face one hurdle for faster growth: ever escalating labor costs. When the standard of living increases, labor automatically becomes expensive. This is true in all countries both developed and developing. A highly competitive market is always on the lookout for either low-cost labor or automation.

As a complement to the above scenario, there is plenty of low-cost labor available in underdeveloped countries. Some countries also have low-cost skilled labor as well. The standard of living in some countries is so low they can offer services for a fraction of what it costs in a developed country.

When we look at these two scenarios, it is not very surprising these countries found what they wanted. While developed countries find low cost labor to produce goods and skilled labor to improve technology, the underdeveloped countries find great opportunities to grow their own economy. Many underdeveloped countries have transformed into developing countries. And when the standard of living in an underdeveloped country improves it becomes a customer in the global economy, increasing the global market size and thereby benefiting the rest of us.

Outsourcing may appear to be one-sided on the surface but in reality it is a bi-directional transaction. For instance, many American and European automobile companies are already selling cars in China, India, and other developing countries. American food chains also operate franchises in these countries. It is also a huge market for cell phones and other electronic gadgets. Construction is booming, with most developing countries using American and European expertise, from architecture to design. Even the international tourism industry has improved as people in developing countries have more money now to spend on their vacation. These developing countries have increased the size of the global market to such an extent the United States and European companies now have to produce more goods and services to satisfy this expanded market. You may call this reverse outsourcing.

However, the benefits of outsourcing will not come immediately. If jobs are outsourced now, it takes time for these underdeveloped countries to start contributing to the global market. When jobs are outsourced to developing countries, several other supporting industries are created within those countries. These webs of industries bring up the standard of living of local people who in turn buy more products increasing the size of the global market. Further, more industries emerge to support this new demand. This *positive feedback loop* keeps expanding and makes the global economy stronger.

In the future, as we begin to see more developing countries around the world, the global economy will be much more powerful.

Dangers of Rapid Growth

Globalization does not come without problems. Economic growth in developing countries happens so rapidly that these nations experience an economic jolt or boom. The scenario is entirely different in developed countries where industries evolved over several decades and there was enough time for overall economic growth.

On the other hand, developing countries are now flooded with industries and investments. Multinational companies have established billion dollar empires. Real estate values have grown exorbitantly high in a very short period because of a surge in demand. Preexisting poverty and newfound wealth do not go together very well. This is not steady growth like developed countries experience; this rapid growth only occurs in some regions of the country. The rest of the regions are still very poor. The difference between rich and poor can cause social and political unrest. Corruption often shows its ugly face.

Globalization is a process of moving up the underdeveloped world. It is a very chaotic process. Though this problem will ease as poor citizens' standard of living steadily improves in these countries, it is a time consuming process if we let it progress on its own.

For instance, there is a great demand for low-cost skilled labor in developed countries and the developing countries can supply it, as their populations are large and waiting for opportunities. The problem, though, are slow and expensive education systems. It takes decades to open a school or a college in a remote, backward village. Even if schools and colleges open, the education path itself is very slow. By the time a student earns a college degree, he or she has spent years learning many irrelevant subjects. These countries must focus on accelerated and low-cost education rather than traditional paths, so their standard of living improves more quickly.

Electronic currencies are very transparent and discourage people from breaking the law. However promising this concept looks, abolishing paper currency is not easy, as people still use it for daily transactions. Abolishing paper currency prematurely will only hurt the economy but the end of paper currency will bring an end to many ills of the current age. How quickly this transformation happens depends on the pace of globalization which again dependent on pace of education system. To make the world economy stronger, pace of education is paramount.

Using This Reasoning to See the Future

In a nutshell, globalization is all about involving more people to pull the weight. If a product or service seems expensive, consider if it has a global market. If not, consider what bottlenecks are keeping it from going global. Once we under-stand the bottlenecks, we will be able to find solutions. The best example is our healthcare system.

In the United States alone, billions of dollars are being spent every year on medical research and development to improve healthcare technology. We cannot stop technology from growing, and every time technology grows, healthcare costs increase proportionally. However, our market size is limited to our population. Without globalization to involve more people to pull the weight, the collected revenue of the healthcare market will be limited. All this investment will come back to haunt us in the guise of increased healthcare costs.

In the future if any new device or technology comes into market, ask yourself if it can be globalized to earn more rev-enue. That is true futuristic thinking. If globalization is prop-erly in place then every time a new technology in healthcare comes into market, the cost of healthcare should decrease. Global demand is a very strong force, we need to identify what is stopping them from going global.

CHAPTER 6

THE INFLUENCE OF AUTOMATION

Understanding Automation

Automation is inevitable as people desire higher efficiency beyond their capability. We already feel its evolutionary effect on our daily lives via ATM machines, self-checkout counters, airport e-check-in, card scanning turnstiles on trains and buses, online banking, and e-commerce to name a few.

Before the invention of ATM machines, we literally had to wait for banks to open on business days to withdraw cash. ATM machines allow banks and businesses to make more transactions and move the economy faster. Just by eliminating the human element from one service, the whole business community gets a boost to move forward at a higher pace.

Another example of automation that enables banks to do business at a faster pace is online banking, which is now so huge that if online banking closed down forcing people to do their transactions in person, banks simply could not handle it. When computers were introduced in banks there was a lot of resistance from employees, who worried automation would replace them. Instead, over the years this automation has created more jobs and improved the pace of the economy to such an extent that there is no going back from this level of automation unless we want to slow down our economy. We bring automation into our lives out of a desire for higher efficiency, faster service, and lower cost.

Industries such as bottling plants and packaging companies have automated most of their jobs. For precision jobs where accuracy is critical, automation is the only answer. Car manufacturers are using robots for precision assemblies.

If automation is helping our economy, why aren't all industries fully automated? Economics. A highly-skilled labor force is required to research, develop, and build technology, which makes robots and machines very expensive while manual labor remains cheaper. It is highly unlikely a costly automated machine will replace a low-cost worker until the machine is mass produced to meet substantial demand for the products or services it produces. For instance, currently there are GPS-controlled, fully automated lawn mowers that can maintain a yard without the need of a human worker. But they are expensive so until mass production brings their cost down, homeowners will keep hiring human gardeners. The same argument goes with robotic vacuum cleaners, and many other home improvement products. As long as there is cheap labor available to do those jobs, automation will happen very slowly.

The Effect of Globalization on Automation

Automation happens from a positive feedback loop of globalization. There are two major factors that put pressure on automation. When industries find expensive labor, they naturally start looking for automated solutions. The other factor is 'increased economic activity' that puts pressure on human efficiency. We already see people working harder to meet deadlines. This makes human labor gradually less effective, forcing us towards automation. Less effective should not be construed as less efficient. It merely means that manual labor will reach its limitation. As labor costs become relatively expensive, it makes automation economically viable.

For instance, scanning technology recently automated many toll roads, eliminating less effective manual labor. This saves travelers a considerable amount of time. After retraining, the same employees can sit at computers and manage this automated toll road system or build supporting software and hardware.

Airline customers can now print their boarding passes online well in advance, saving time at manual check-in counters. Fast checkout lanes are examples of automation in supermarkets.

Wherever you see more economic activity and higher labor cost, automation is showing up. High-paced economic growth demands automation. Globalization feeds high-paced economic growth in all parts of the world.

The other motivation to replace human element is that, the quality will be *consistent machine quality* because of automation. Increased production rates mean speedier deliveries, winning the industrial production race against other competitors. This higher market share quickly compensates for the large investments made in the automation. Remember: automation in any industry is very cost sensitive; it will not happen unless the market demand is huge. Only an extensive globalization can initiate automation.

As the above scenario suggests, automation becomes inevitable in all global businesses when the global economy expands. The massive outsourcing that is happening today is a preceding step before automation. Outsourcing is one of the elements of globalization that is increasing the economic activity.

Automation will be more in developed countries where labor cost is very high, less in developing countries where labor cost is moderate, and little to none in underdeveloped countries where labor cost is cheap. However, as countries develop, they all will go through the same stages of globalization and automation.

Is Automation Making Us Safer?

Automation is the only answer to manmade disasters. We have witnessed bridges collapsing, aircrafts crashing, and tragic offshore oil leaks. In the case of bridges, for instance, engineers may have unintentionally overlooked some parameters—maybe a wrong material, maybe more load than expected, or simply poor design or maintenance. If machines fabricate and manage bridges instead of humans, there will be more consistency and fewer errors. If there is a mishap, it is easy to identify the problem and correct it in the next enhancement. The same argument goes with every other man-made disaster.

To achieve higher levels of safety, we need to move into higher levels of automation. Globalization will make this happen at a higher pace.

Effects of Automation on Future Jobs

We will see considerable job losses during the early stages of automation and equally massive job creations in new areas. We are already seeing such overlaps. Companies like Google, Apple, Facebook and Amazon were unknown a few decades ago. Today those companies employ thousands of people all over the world. Looking at job losses from an evolutionary angle requires we look at both sides of the coin. We are losing jobs in some areas where manual labor is less effective yet creating jobs where automation is needed.

There is no one-to-one relation between job losses and new job creation so it is important to look at overall economic development rather than any one particular company where jobs are lost. For instance, if the state lays off highway tollbooth employees, they will be hiring others to manage the automatic systems and will also need to contract other companies that provide computer networks for faster connectivity, hardware suppliers, and a host of other infrastructure needed to support the business. Similarly, if a grocery store lays off checkout cashiers in favor of self-check machinery, the company that designs and programs the machines and all supporting businesses will be hiring employees.

Our job market is changing so fast that a person who loses their job may find a similar position is not available. Instead, they need to look for industries that are creating new jobs. If knowledge or training in IT is lacking, then retraining is required.

Here is a classical example to show how a growing economy inevitably sheds manual labor. The furniture in our homes, for instance, used to be handmade. Carpenters cut the wood, prepared the surface, and assembled the final product by hand. Now, high-precision, computer-controlled machines cut and prepare the wood with humans only handling the final assembly. This improvement in technology is the reason modern furniture is becoming less and less expensive. Compare a wooden chair that costs only $10, to one manually built by a skilled carpenter paid $50 or more an hour. How many chairs must that person make in an hour for the company to make a profit? That business cannot survive without automation. Carpentry skills are not needed for mass production purposes. The skill needed is the ability to manage machines.

This does not mean there are no carpentry jobs any more. Today's carpenters are largely focusing on customized work, which is not mass-produced, like custom cabinets, hand crafted furniture to name a few. Even in these jobs, they are faced with automation: machines prepare most of the wood used. Even the final assembly is done with automated tools to save time. If you closely look at the history of any profession, you will see a pattern of continuous automation.

In this fast changing world, anyone who loses a blue-collar job and finds it difficult to secure a similar job must realize that the world has moved on and seize the opportunity to change careers.

Process by process, job by job, automation is creeping into our lives, improving our lifestyles. This automation blends so seamlessly into our lives people many times do not notice it. It makes our lives richer and many times gives us no other option. Life has become very hectic and stressful, making the need for these technological products even more necessary, with less and less time to perform such functions ourselves. This is why automation is irreversible.

Eventually automation will have profound effect on everyone's career. Many people have questioned the ethics behind automation when it is going to cause extensive job losses. Here is an example to show how we embrace automation without our conscious knowledge. Suppose there are two companies selling the same product for the same price including shipping and handling. Company A is fully automated and delivers the product in two business days, while company B is not automated and only uses human labor and delivers the product in five business days. Which one will customers prefer? Non-automated companies cannot compete with automated ones. They might to some extent when the labor cost is low. However, when the labor cost increases, it becomes hard and eventually they loses out to the competition that uses automation. Most businesses face an ultimatum of automate or give up as manual labor cannot produce goods at machine speed or cost. Speed, quality and cost will be the winner and will become one of the major evolutionary criteria in all the industries.

What does the real automation look like?
In the real world, there are many automation processes that we can pay attention to, especially in our own eco system. Our eco system is a highly sophisticated, matured, and self-sustaining automated system. Consider honey bees. They travel from flower to flower, collect nectar, and create the honey. They live on their own without any human intervention.

Imagine the cost of honey if there were no honey bees and humans had to mimic their nectar collection process.

Plants and trees provide fruits and vegetables for us using a highly complex system that synthesizes food cell by cell. Imagine the cost of food if humans had to mimic their process.

Almost all the natural processes happening around us are fully automated. They are highly complex, self-sustainable, and free for humans. That is what the automation process should look like when fully developed. The baby-steps of automation we are developing today are trying to work toward: mimicking Mother Nature. It is needless to say that a highly skilled labor force is needed in the field of automation.

As we progress into the future with better technology, it puts pressure on the education system. This is where our futuristic eyes should focus on.

CHAPTER 7

THE INFLUENCE OF TECHNOLOGICAL EVOLUTION

Understanding Technological Evolution

When people spend money to increase their living standard, they not only grow the economy but also drive technological evolution. Here is the distinction between economic growth and technological evolution: A car company expanding to produce more cars is considered economic growth. If the same car company manufactures cars that are more efficient, it is technological growth. When we buy a car, we promote the economy; however, when we shop for a more efficient car, we promote better technology.

The economic growth drives the technological growth and vice versa, and our desires drive both. Without human desire, there is neither economic growth nor technological growth.

Which is more important: sophisticated technology or strong economy? The truth is, one can't be better without the other. Suppose engineers design gigantic tidal barriers to protect coastal cities from hurricanes caused by global warming. Only an equally strong economy could handle such construction. If our economy is weak, all our technological marvels will remain on the drawing board.

To understand the technological evolution, we need to understand the implicit relationship between globalization and automation. Globalization increases the economic activity of the world, creating a need for automation. Automation on the other hand, makes it easy for the globalization to happen.

For instance, washing clothes was originally a manual service and difficult to globalize. Then technology developed and washing machines were developed, which quickly became a global product and is now sold in every big box store. This local service was globalized because of automation. Further, the same globalization brings in more automation to produce more and better machines.

The same is true with dish washing machines, microwaves, toasters, and a host of friendly machines around us. When a process is automated it becomes a global product.

Without automation, globalization would be exceedingly slow. The telex and telegram systems we had in the past are all examples of early stages of automation that were very slow. Now, with faster communication networks, the speed of globalization has picked up. We cannot imagine outsourcing so many jobs without these faster networks. The pace of automation decides the pace of globalization.

These two human endeavors—globalization and automation—together drive our technological evolution and make our global economy stronger. One cannot succeed without the other. It resembles a seesaw, where a single child cannot play alone.

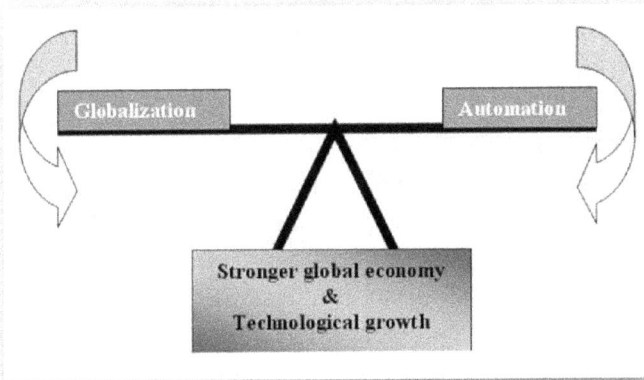

Fig. 5: Globalization and automation, the two playmates

While globalization represents the quantitative aspect of an economy, automation represents the qualitative aspect of an economy. Both quantity and quality matters for the technological evolution. If we build a huge bridge using manual labor, we only help build the economy. However, if we build the same bridge with sophisticated tools and machines, we progress technologically.

The Dependency Shift
While human desire is labeled as the driving force behind technological evolution, the strategy that we are employing to facilitate this evolution is to move away from human labor, eco- dependence, and obsolete technology.

We can call this strategy the *dependency shift*. The Industrial Revolution of the 19th and 20th centuries was just the beginning of a massive technological dependency shift from manual labor towards machine. Inevitably, every one of us has embraced this dependency on machines in our lives, as it satisfied many human desires.

Automation and Dependency Shift

Automation is comprised of multiple smaller dependency shifts. For example, if a screwdriver is redesigned with a better handle so that workers can work more efficiently, it is a minor dependency shift. However, if a machine is invented that can do the same job at a much higher pace with the push of a button, this is a significantly greater dependency shift. Whether it be through several smaller dependency shifts or a single large one, the end goal of the dependency shift is the same: full automation. When you envision solutions to current day problems, you have to envision all possible dependency shifts. Many times smaller dependency shifts are needed to suit the buying capacity of the consumer. For instance, self-driving car technology is evolving and is still expensive. Now to suit the buying capacity of the market, car manufacturers are coming up with smaller automations like break assist, crash avoidance, blind spot monitoring, and automatic breaking systems to name a few. Eventually they all will converge into one system, which is the self-driving system.

Globalization is all about spreading the dependency shift across the globe. When businesses go global, technology will grow faster because of the mutual exchange of resources like energy, materials, skills and so on. Businesses around the globe are achieving dependency shifts in smaller chunks. It is just a matter of time before we achieve a life of zero dependency and total automation.

It is hard to determine what kinds of technology is sufficient in achieving this, but that is irrelevant because neither the consumer nor the producer care what technology is the best as long as it satisfies the needs of the consumer and produces profits for the producer. For example, there are a variety of cellphones utilizing different operating systems and wireless technologies, but all the consumer cares about are the speed, reliability, and interactive features on the device, and the manufacturers simply focus on these evolutionary criteria using whatever technology they own at that time.

The surprising fact in all these evolutionary phases is, survival from extinction-level threats will never become the driving force. The driving forces for future technological evolution will always be comfort, luxury, and our desire for better living, which makes us technologically stronger day-by-day and eventually capable of surviving extinction-level threats.

As we enter the Age of Automation, we will become stronger and more capable of surviving global disasters. Our technological evolution will move us at a faster pace going forward as machines do most of the jobs.

Researchers have been unearthing intriguing inventions like gears, clocks, even computers and airplanes dating back several thousand years and have wondered how someone from an ancient civilization could have envisioned them.

If we understand the driving forces of technological evolution, there is nothing to be surprised about such inventions. All civilizations have inventors. Just as we dream about our future, our ancestors dreamed of their future, which is our present. But they were lacking an advanced economy to support such advanced technology. A great invention needs an equally powerful economy.

If the economy remains healthy, then technological advancements happen continuously. If the economy is weak, no matter how grand our ideas and how deep our desires are, they will be just dreams.

To grow technologically stronger and faster we need to keep our economy strong so any discussion on technological evolution is incomplete without addressing the forces that are responsible for economic recessions.

CHAPTER 8

THE INFLUENCE OF ECONOMIC RECESSION

Understanding Economic Recession

Humans have both the desire to spend and the desire to conserve. While these characteristics may seem at odds with one another, they are two sides of the same coin. When people spend the economy grows, and when they conserve the economy slows down. The reason people spend and the reason they conserve decides the fate of the economy.

People become conservative and buy or invest less for various reasons. Losing money in the stock market and real estate are well-known reasons for people to become conservative. When people reduce their normal spending, it affects the economy. When spending slows, industries produce less and may lay off people. Small businesses that cannot absorb this slowdown declare bankruptcy. Once the economy slows down, more people become conservative in their spending. This negative feedback loop is a self-destructive force, which eventually leads to economic recession. This is why governments spend billions of dollars to stimulate the economy to break this negative feedback loop.

The Economy Can Hurt Itself

The concept of credit was developed to grow the economy faster and reduce the bottleneck of cash availability. Though credit has a very positive impact on our economy, it has a huge negative impact when not regulated. Unregulated, easy credit creates fictitious growth of the economy exposing both producers and consumers to a higher risk.

During high economic growth, easy credit motivates people to buy or invest more, be it a car or real estate or any product of value. There is nothing wrong in using credit; however, if people use credit to buy beyond what they can afford to pay back, they end up creating a fictitious economic world. This means companies produced more goods than they were supposed to produce.

Suppose one million people each spend a thousand dollars shopping in a year, using their available credit. This amounts to one billion dollars flowing in the economy. Companies produced goods and services to meet this demand. It is also possible some new companies entered the market to meet this demand.

Now, let us say the one million people have reached their credit limit. Some responsible customers paid back their loans and continued to use their credit cards, while others who were not capable of repaying their credit card balance, began to use other credit cards, risking themselves and the economy even further. When these risky customers fail to repay and then declare bankruptcy, it not only hurts creditors but also hurts companies that produced the goods and have planned more for the growing economy. They may have hired more people and bought more inventories for future growth. Adding to this confusion, many new industries were established assuming the economy was growing strong.

In the real world, these numbers could be bigger; instead of one million there could be 10 million or more risky customers involved in this reckless spending—even more globally. In addition, the spending amount could be as much as $10,000, $20,000, or more per person. This amounts to billions or trillions of fictitious dollars floating in the economy.

Now, when these risky consumers reach their credit limit and are no longer part of the potential market, companies will suddenly experience slower sales and their profits plummet. The money they will have invested to produce more products is now at stake. Eventually they start laying off employees, which will have ripple effect on the other industries. This is the beginning of a negative feedback loop, which may initiate an economic slowdown. Once this loop forms, it continues to slow the economy. Credit could be devastating if not regulated.

The same easy credit causes bubbles in the real estate market too. Remember, real estate value depends purely on demand. If fictitious demands stemming from easy credit drive up the market, bubbles are bound to happen.

The easy credit devil is right behind any economic boom to bite it if not regulated. Regulatory agencies' are blamed for their inability to decide how much to regulate and when to regulate. The volume of available financial data in the world is growing at an astonishing speed and it is difficult to handle it without well-developed software. Until our IT catches up to the needs of our fast-changing world, regulation is not an easy task.

IT is our window to the financial world. It has catapulted our economy to this level and only it can control its growth. The higher its sophistication (clarity), the better our view of the financial world will be. When IT lags behind the changing world, it will create unpredictability and harm the economy. It does not matter at what evolutionary phase we are. A closely monitored economy is the fastest way to progress.

From an evolutionary point of view, every recession keeps us technologically weaker. When our technological evolution pauses, every day we are losing the opportunity to move closer to the Age of Automation. Below is an example to illustrate how massive a recession could be.

The Economic Recession of 2008

The stock bubble, housing bubble, and energy bubble are the three undesirable forces responsible for any economic recession. Even though each one of these events is capable of causing serious economic damage, when they hit us at the same time, the effects are felt worldwide. This is exactly what happened during the 2008 recession. When the housing market collapsed in 2006, risky credit began to hurt the stock market, which stumbled in 2008. Oil prices were then at their peak as the economy was booming. This energy-inflicted inflation began to hurt the real economy even more, causing the inevitable recession. This is what I refer as the *meeting of the three devils.*

The Housing Bubble

In the graph below, we see prices have steadily risen during 2006 that resulted in a market crash. This housing bubble affected the entire world during 2006.

With the real estate market becoming global, people are not limited to their own country for investments. It is very common to see people residing in one country while buying property in other countries for investment purposes. Real estate entrepreneurs have invested in multiple countries. In a global perspective, we see a common trend in the real estate market in most developed countries, where a rise or decline in price affects investors in several countries at the same time. Whether these price adjustments cause a market crash will depend on the number of people affected and their capability to survive the fall. The more their survival capabilities are the less impact the crash will have on the economy. However, if it affects the standard of living for a significant number of people, then it becomes a real market crash.

If we can find out why there was a sharp rise in price, we can create a solution for such market crashes in the future. In upcoming chapters, we look at the real estate market from the evolutionary perspective to find real world solution.

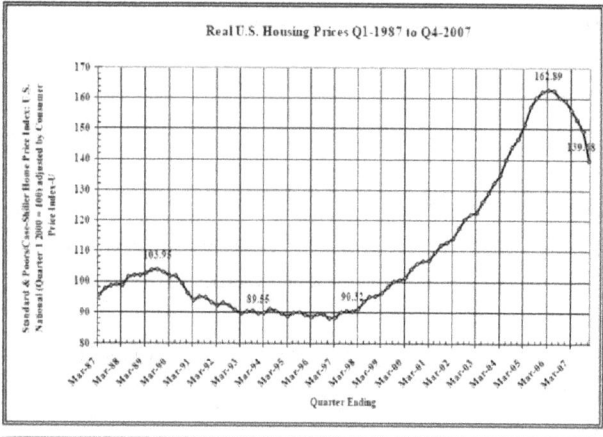

Fig. 6: Real US housing prices Q1-1987 to Q4-2007

The Stock Market Bubble

The graph below shows the performance of the major global stock indexes from 2000 to 2009.

Fig. 7: Major stock indexes from 2000-2009

The Energy Bubble

The graph below shows the energy prices at their peak before the recession began.

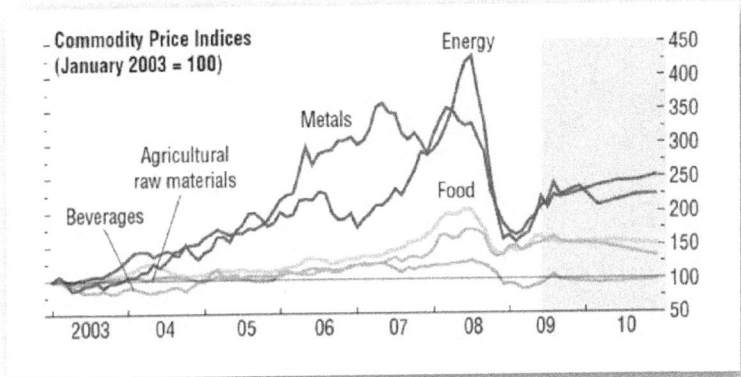

Fig. 8: Energy and commodity prices 2003-2009

Although many economists did spot these bubbles, they failed to understand their severity. These bubbles exist because of a lack of information flow in the system coupled with our inability to regulate the market. We may not be able to change human nature, but we can regulate our economy appropriately, not by imposing rules and regulations that can be broken, but via automation using information technology.

PART III
A GLIMPSE INTO THE FUTURE

CHAPTER 9

THE FUTURE OF HEALTHCARE

I n order to adapt and evolve we must remain healthy as individuals and as a species. Unfortunately, the current state of our healthcare is in peril, as costs increase every year.

According to the United States Department of Health and Human Services, the total healthcare spending in 1970 was about $75 billion, or only $356 per person. In less than 40 years, these costs have grown to $2.2 trillion, or $7,421 per person.

It is not only in the United States that costs are increasing, but also in every developed and developing country. However, the rise in healthcare costs in the United States far exceeds that of other countries.

The chart below shows the Healthcare Spending in U.S since 1965.

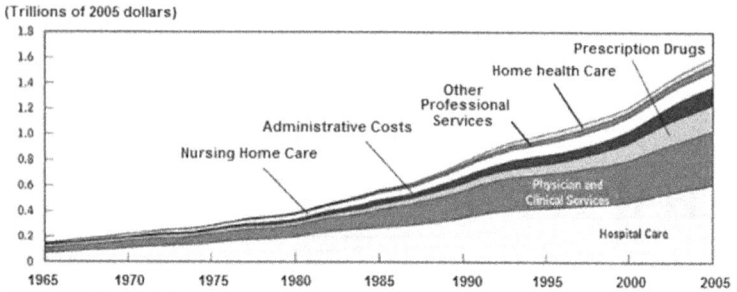

(Trillions of 2005 dollars)

Prescription Drugs
Home health Care
Other Professional Services
Administrative Costs
Nursing Home Care
Physician and Clinical Services
Hospital Care

Fig. 9: Real spending on health care in selected categories, 1965 to 2005

The experts' explanation of rapidly increasing health-care costs is people are getting more care, much of which is associated with new medical technologies. Some analysts say the availability of more expensive state-of-the-art drugs and technological services fuels healthcare spending. Not only is this because the development costs of these products must be recouped by the industry, but also because they are generating consumer demand for more intense, costly services, even if they are not necessarily cost effective.

The nature of healthcare in the United States has changed dramatically over the past century, with longer life spans and better care for people with chronic diseases.

The data below shows an increase in the usage of some of the latest technologies over time in the United States.

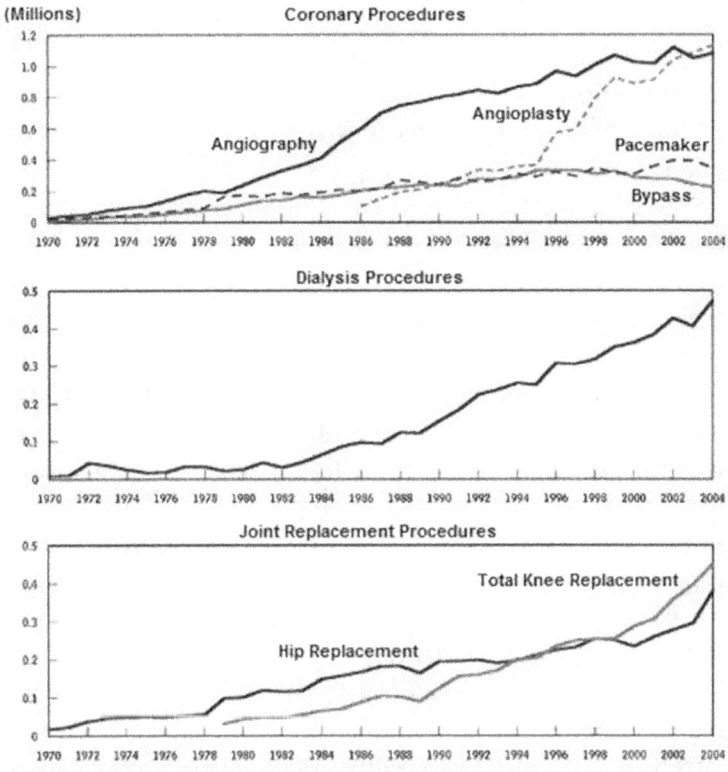

Fig. 10: Use of selected health care procedures by people aged 50 or older, 1970 to 2004

It is estimated that healthcare costs for chronic disease treatment accounts for over 75 percent of national healthcare expenditures. According to the Centers for Disease Control and Prevention, chronic diseases are the leading causes of death and disability in the United States. Healthcare technology advances every year. Advanced medical technology diagnoses more and more people with new problems every day, increasing the number of patients receiving treatment.

As recently as a few decades ago, diagnosing cancer was relatively rare. Now, doctors diagnose many types of cancers on a regular basis. Many argue that modern habits cause cancer in more people. Though true to some extent, the reality is, we simply did not have the tools or technology to diagnose it as easily in the past. Once diagnosed these diseases have to be treated.

Clinical developments continue to increase the longevity and improve the quality of life of many Americans — gene therapy for cancer treatment, less invasive surgical procedures, robotic surgery for greater precision, advances in reproductive technology, infectious disease control, and so on. The amazing world of nano-technology is unravelling many nano-healthcare options. All these new technologies add to the cost of healthcare.

This sophisticated medical technology results in expensive treatments and is a major cause of higher healthcare costs today. *There is a direct relation between the technological developments and healthcare costs.*

Health insurance premiums continue to grow as healthcare cost keeps climbing. Many people argue private insurance companies are the cause of these higher costs of healthcare. However, to some extent, they are responsible for the higher costs because of their own inherent overhead management costs, even after making these insurance companies more efficient (automation will make it happen eventually), increasing costs will not go away until the root cause of the problem—technology—addressed.

Fewer Tests to Bring Down Cost?

The current trend among many physicians is to order as many tests as presumably relevant to make more accurate diagnoses and to protect themselves from malpractice lawsuits. This surely increases healthcare costs.

Many healthcare professionals argue that doctors could understand illnesses better by spending more time with patients. That would help them order tests that were genuinely needed and eliminate unnecessary ones.

More tests may seem unnecessary, but will help in making an accurate diagnosis and many times will unearth some hidden problems the patient never even knew about. Many tumors and other diseases may not show any symptoms until they become life threatening. Many times these conditions are discovered while investigating some other health problem. More tests are always life-saving.

The Evolutionary Solution

Just like any other industry, if a product were mass-produced or if a service were mass supplied, the cost will decrease. As we have learnt, globalization is all about involving more people to pull the weight. In the United States alone, we are investing trillions of dollars on healthcare technology, but our market size is limited to our population. Although some healthcare companies are exporting medicine and healthcare equipment, they still have a long way to go. Without globalization, the collected revenue of the healthcare market are limited and obvious result is the raise in healthcare cost.

In a similar fashion, computers were very expensive when the market was small, but became very affordable despite its technological sophistication with a global market. In 80's the cost of a PC was more than $2000 with 3Mb memory and 30Mb storage. Today, that old super computer that would have set you back over $20 million would barely be able to run Windows XP. It is merely a numbers game—the bigger the market the cheaper the product. Globalization is the 'silver bullet' for reducing American health-care costs.

Size of Global Healthcare Market
World population has exceeded 6.5 billion and is growing.

WHO region	Total health workforce		Health service providers		Health management and support workers	
	Number	Density (per 1000 population)	Number	Percentage of total health workforce	Number	Percentage of total health workforce
Africa	1 640 000	2.3	1 360 000	83	280 000	17
Eastern Mediterranean	2 100 000	4.0	1 580 000	75	520 000	25
South-East Asia	7 040 000	4.3	4 730 000	67	2 300 000	33
Western Pacific	10 070 000	5.8	7 810 000	78	2 260 000	23
Europe	16 630 000	18.9	11 540 000	69	5 090 000	31
Americas	21 740 000	24.8	12 460 000	57	9 280 000	43
World	59 220 000	9.3	39 470 000	67	19 750 000	33

Fig. 11: Global health workforce by density[1]

The data from the World Health Organization (WHO) indicates that in the African region, there are only 2.3 physicians per 1,000 people, compared with 24.8 per 1,000 in the American regions.

---

1 *All data for latest available year. For countries where data on the number of health management and support workers were not available, estimates have been made based on regional averages for countries with complete data.*

The majority of the world population is poor. The actual size of the global healthcare market depends on the cost of the healthcare. If the cost is high, only a few can afford it; however, if the cost is low, a significant population can afford good healthcare. Cost and market size always work against each other. Most products and services start with high cost that only rich can afford and gradually become cheaper as the market grows. If the healthcare market goes global, it will be no different.

Globalization is the 'silver bullet' for reducing healthcare costs. Few globalization scenarios are discussed below.

Globalizing Diagnostic Services

With the abundance of websites and apps now offering online medical information, patients are increasingly self-diagnosing themselves when they experience a new health problem. That can be alarming for doctors when patients show up with a scary list of unlikely scenarios, demanding expensive and unnecessary tests. More troubling is when patients fail to seek expert medical advice after misdiagnosing their problem. The healthcare industry has responded to this trend in a proactive way. Many physicians are offering computerized systems that can provide reliable medical information or diagnoses. These services are also incorporating electronic medical records (EMR), encouraging patients to use them before visiting their doctors to save time and make consultations more productive.

An EMR is a digital version of the patient's medical history a doctor keeps on file. With EMR data, medical providers can respond quickly to deliver the best possible care. Patients are less likely to undergo redundant tests and physicians don't lose time waiting to see test results from a lab.

Electronic health records (EHRs) on the other hand are inclusive of a broader view of a patient's care containing information from all the clinicians involved in a patient's care. Broadly speaking, the EHR format is geared more towards sharing across multiple organizations or across countries.

Both EMR and EHR are still evolving. However, they are already affecting the way patients are diagnosed and treated. The challenge is that as a patients' EMR data keep growing—the data within EMRs can span a lifetime of clinical notes, lab results, and medication history—the resulting huge database will create a challenge for doctors needing to sift through it, especially during emergencies. Fortunately, another technology can solve this problem.

After years of research at Cleveland Clinic, IBM Research has unveiled Watson, a cognitive computing system that interacts with physicians and uses patients' EMR and medical literature to help diagnose more accurately in real time. It will not only help physicians to focus on key information from enormously large data banks but also from medical journals, clinical guidelines, trial results, etc. Watson uses machine-learning methods to digest tremendous amount of data, and with feedback from doctors after each diagnosis, the system diagnostic abilities get that much better every next opportunity.

According the published specification, Watson can digest more than 600,000 pieces of medical evidence, more than two million pages from medical journals, and search through up to 1.5 million patient records. For a human doctor, it would take hours of reading just to keep up with new medical knowledge as it's published. These mismatches will widen exponentially as technology progresses in the future. It's just a matter of time before the doctor's presence become less relevant for routine diagnosis and more relevant for developing online diagnostic systems.

This paradigm shift is truly helping globalize healthcare. One great benefit to this evolutionary adaptive healthcare system is that the world becomes the target market. Anyone, anywhere with an Internet access will be able to use an online healthcare system and be diagnosed. As no human doctor is physically involved, the cost will be very affordable. Diagnosing most routine healthcare problems digitally not only substantially reduces healthcare costs but also increases the revenue considerably. Doctors can focus more on research to make the system much more productive.

Patients in some developed areas of the world will not need to key their medical history into such a system, as most of their medical information is already available online. However, patients from underdeveloped regions of the world may have to input their entire medical history at least once before diagnosis. Once sent, this information is stored online and updated when the patients next use online diagnosis. As more and more diagnoses and treatment information goes into the database, the system becomes more robust.

Globalizing Surgical Procedures

Significant advances in automation are happening in the field of robotic surgery, a sophisticated platform designed to enhance the surgeon's capabilities and offer minimally invasive options for surgeries. The surgeon uses very small tools attached to a robotic arm controlled by a computer, enabling many types of complex procedures with more precision, flexibility, and control than is possible with conventional techniques.

Traditionally, surgeons needed to make large incisions to reach the areas they were operating on. This resulted in long recovery times and readmissions. During the 1970s laparoscopic technology emerged and revolutionized the surgical field with minimally-invasive surgery. Robotic surgery builds on that advance.

There are many companies competing in the field of robotic surgery, each focusing on evolutionary criterion such as flexibility and accuracy of the surgical instruments, better touch and feedback for the surgeon, and better visualization of the patient's anatomy. Some robots like Telelap are also focusing on eye-tracking technology, which allows surgeons to manipulate and control the instruments by gazing at various parts of the screen.

Because many of these models are still in research and development stage, the real arms race has yet to begin. For now, Intuitive's Da Vinci's surgical robot is well-established in this market. It is manipulated from a remote console via hand controllers and foot pedals, enabling surgeons to perform an array of activities including pre-operative planning and surgical rehearsal, open and minimally invasive surgery, remote tele-surgery, and image-guided surgery.

Although there are numerous complaints and malpractice law suits against these robotic procedures, if you look at the real statistics, there is no doubt robotic surgery is the technology of the future. Here are some interesting facts.

- The first robotic surgery occurred in 1997.
- The first robotic coronary artery bypass was performed in 1999.

- In 2000 the da Vinci Surgical System became the first robotic system in the United States to be cleared by the FDA.
- More than 1.7 million robotic procedures were carried out between January 2000 and December 2013.
- The first unmanned robotic surgery was performed in 2006.
- Most robotic surgeries need only two or three small incisions to complete a surgery.
- According to studies, blood loss during operation, hospitalization, and readmission cases are much lower with robotic surgery.
- The rate of robotic surgeries is increasing by 25 percent annually.

Although robotic surgery is not yet ideal for all procedures, it has replaced traditional open surgery for prostatectomies, nephrectomies, hysterectomies, and cardiac valve repair. There is a great potential for globalizing this field to bring down the cost of healthcare.

With faster Internet and higher bandwidth, the future of telemedicine is very promising. Telemedicine increases global competition among hospitals and doctors and technology companies to offer their best. Consumers are the pure winners in this game.

M2M Technology to Reduce Hospital stays
The growing elderly population around the world is driving the need to monitor patients remotely. The adoption of

wireless technology in our personal lives has accelerated the growth towards mobile health monitoring.

Today, one of the core drivers of healthcare industry is Machine-to-Machine (M2M) technology. M2M technology or telehealth utilizes the pairing of medical devices and communication technology together to remotely monitor diseases and symptoms in patients.

With increased acceptance of M2M, the hospital occupancy rates continue to slowly decline. This not only decreases the healthcare cost, it's a great relief to elderly and patients with chronic health issues for whom recurrent hospitalizations are a major source of distress.

In 2015 alone healthcare providers remotely monitored more than 4.5 million patients worldwide for diabetes mellitus, hypertension, congestive heart failure, chronic obstructive pulmonary disease, and mental health conditions, according to a global health report. They are expecting this number to grow as the devices become available globally, further reducing the hospital stays and healthcare costs.

Upcoming Technologies Helping Further Globalization

Nanotechnology in medicine is another highly promising technology that is capable of performing repair at the cellular level. These advances could revolutionize the way we detect and treat diseases in the future, and many techniques only imagined a few years ago are making remarkable progress towards becoming realities.

One application of nanotechnology in medicine currently being developed involves employing nano-robots or nanobots to deliver drugs to specific types of cells, such as cancer cells. These are engineered so that they are attracted to diseased cells for treatment. This technique reduces damage to healthy cells in the body and allows for earlier detection of disease.

It is expected that the global value of nano-enabled products and nano-materials could reach as much as $4.4 trillion by 2018.

With all these automations converging, healthcare will inevitably globalize, reducing the cost of healthcare because of the market's sheer size. Any further reduction in healthcare costs through mass-production of devices, initiates the positive feedback loop. As the world's standard of living increases, more money will flow into the healthcare industry and create solutions where there were once only problems.

What we are trying to achieve here is a paradigm shift from physical consultation to an automated healthcare system with built-in intelligence to diagnose and treat patients remotely in real time with more precision, lower cost, and at unbelievable speed.

With your futuristic cap on, what kind of future do you see in healthcare? With Watson reading medical literature for diagnosis, robots doing surgery, and nanobots monitoring you 24/7, in my opinion the field of medicine is slowly becoming a field of machine-building to treat patients, reducing the need for a human doctor. I believe this technological development is inevitable because they offer more precision, unstoppable because of huge market demand, and irreversible because they offer cutting-edge technologies. Every time a new technology emerges, it either helps existing technologies grow faster if they are relevant or makes them obsolete if they are otherwise. As futurist we need to analyze these mutually influencing technologies while drawing a timeline of future events.

Future of Malpractice Insurance

Medical malpractice insurance is in place to maintain high-quality healthcare. Doctors and hospitals pay huge premiums to cover any human error that results in a wrong diagnosis or treatment regimen that may eventually have severe, damaging, or even fatal results. This cost trickles down to the patients, resulting in higher healthcare costs for all involved.

A good medical diagnosis depends on precise lab tests. While more lab tests lead to higher healthcare costs, fewer lab tests can lead to the wrong diagnosis, resulting in exorbitant legal costs and malpractice claims. Either way, patients are the ones penalized. Automation in healthcare system is a simple solution to this problem. Automation eventually reduces the cost of malpractice insurance resulting in further reduction of healthcare costs.

Automation in Healthy Living

Whether we eat in a restaurant or buy food in a supermarket, for the majority of the population, cost is one of the driving factors when deciding what to buy, when, where, and even how often. Since it is often more expensive to eat in a sit-down restaurant—particularly for a family of four, for instance—junk food and generally unhealthy choices often win over healthy food based on cost. Junk food is also often tastier than its healthier counterpart so there is no incentive to choose a healthier food alternative.

For example, a bag of chips costs a couple of dollars, while the same amount of fresh salad costs two to three times that amount. This bottom-heavy analysis of healthy food versus junk food has literally changed the food habits of an entire generation of humans, resulting in obesity, diabetes, and other long-term health issues at very young ages around the world.

It is also not always possible to keep track of the calories in the food we eat. The effects of over-eating are slow to manifest themselves, and many only notice a health condition like obesity or diabetes or heart disease after several years— or even decades. When these people visit doctors and receive treatment in a hospital, the cost of healthcare goes up.

The solution is real-time health monitoring that can easily find solutions to these types of medical issues. With automation in health monitoring, we will soon wear monitoring devices to test our blood and monitor our food intake. Affordable, simple, and practical devices like blood sugar monitors, for example, allow people to get immediate feedback on their blood chemistry. As this approach is being promoted by insurance companies, we would find a huge demand for these machines and companies competing to produce state-of-the-art products to help people monitor their blood chemistry with as little discomfort as possible.

For certain people with diabetes, the biggest innovation may be Dexcom's Share2, which displays glucose data on a watch or smartphone. A Dexcom sensor with a hair-thin wire is placed just under the skin that sends glucose data via Bluetooth to an iPhone. With diabetes on the rise, this technology will grow fast as the market size is huge. We could expect more investment to go into the monitoring of blood chemistry, which will advance the technology behind preventive healthcare.

The next wave of smart devices that are coming into the market are in the field of health and fitness. The demand for smart technology and wearables is huge. As of 2015, some of the popular ones are:

- Timex Sport by Ironman
- Forerunner 920XT by Garmin
- TICKR Heart Rate Monitor by Wahoo Fitness
- Magellan Echo Fit Sports Watch
- FitBit Surge
- Nike + Sensor
- TomTom Runner Cardio GPS
- Unnto Ambit3 Sport
- Polar V800 Sports Watch
- Fenix 3 by Garmin
- Mio Alpha Heart Rate Sport Watch
- Adidas MiCoach Smart Run
- Sensoria Technology Smart Sock

Each one of these competing devices are designed to help athletes to do their tasks more efficiently. Using GPS capability most can count step and swimming strokes, compute calorie intake, and so on. Some also have a heart rate monitor, and sleep tracker that analyzes sleep patterns and recovery. A few can even plan a person's training and work-outs and suggest diet plans.

The demand for these devices is so huge that many such technology projects are being crowd funded on Kickstarter. This trend indicates our awareness on healthy living is growing. There has always been a strong desire to live long and be healthy, but now we can track it! As long as this market is strong we will see more money flowing into investment to do more research to bring better products.

Until nanotechnology expands into mainstream health-care monitoring, the current noninvasive devices will continue to evolve, although getting intricate data from human body will always be a challenge. When nanobots begin to monitor the human body, the existing smart devices will reach their tipping point and grow exponentially to capture the market.

Healthy Restaurants

Many restaurants are already listing nutritional information on their menu. In the future this becomes a necessity so people can choose the right food and right quantity to keep good health as well as to please their insurance company.

The devices discussed above will evolve to track restaurant menus. Instead of a hand-held menu, you will check the screen of your smartphone, which will display the restaurant menu along with its recommendations to match your daily nutritional needs based on your current bloodwork, which is stored in the database.

With this kind of forced monitoring, restaurants will be extra cautious in what they feed their customers. Soon insurance companies will control what we eat in an attempt to keep us healthy. Prevention rather than cure will be their business focus.

Further into the Future

The upcoming technologies—robotic surgery, IBM Watson, nanotechnology and a host of smart devices—will greatly influence each other to grow smarter. For example, surgical robots could interact with nanobots for precision. They also could interact with Watson for more patient-specific information. Watson could also become much more efficient with data feeds coming from nanobots. As these devices begin to learn from each other, it leads to the birth of an intelligent healthcare system.

The automation of the healthcare industry speeds up the flow of medical information. Hospitals around the world will have instant access to every new breakthrough, diagnosis, and treatment because healthcare will exist on a connected global network. The machines will instantly assimilate new information as it becomes available. If a new epidemic hits a region of the world, its symptoms and needed treatment will become instantly available to the global healthcare system, so the whole world benefits from this instant flow of information.

With healthcare automation, all the money now spent on routine checkups and meaningless paperwork will go into intellectual research. We will quickly attain a higher health standard in a short period. This concept is at the heart of what holds us back in the current age. Letting machines take over more and more of our routine jobs frees us to stop working so hard and start working smarter—the only move assured to take us into that safer future.

CHAPTER 10

THE FUTURE OF ENERGY PRODUCTION

Solar energy comes to our planet in the form of sunlight. All plants on earth store this energy through photosynthesis. Herbivorous animals get this energy by eating plants. In turn, carnivorous animals eat animals for energy. It would be accurate to say most living beings on earth derive energy either directly or indirectly from the sun, with the exception of some plants and animals that derive energy from chemical reaction.

Why then did our search for energy begin with digging for fossil fuel? Why did not we invest in this great source of energy?

The answer, of course, lies in our technological evolution itself. We can trace our current dependency on non-solar energy back to our very first engine that, as you will recall from history textbooks in grade school, was a steam-powered engine, not a solar-powered engine.

During that time, our knowledge of utilizing solar energy was primitive at best and nonexistent at worst. The world celebrated the invention of the first steam engine because it was the greatest evolutionary product of that time. We embraced the technology and our economy began building on this energy source almost immediately. Capitalizing on this technology, we were able to move many more goods quickly, evolving at a steady rate. Gradually, coal power became the backbone of our economy.

We ate solar food but traveled on coal!

Then came oil—a more sophisticated source—for almost all of our energy needs. In spite of its sophistication, it has not totally replaced coal even today. Coal is so deeply rooted in our economy that to this day there are hundreds of active coalmines around the world.

Literally thousands of coal miners around the world are still working hard to mine the fabled *black gold.* The energy generation from both coal and oil is very well established and both are quite vast sources. The sad part however is neither source support a sudden surge in energy demand.

The alternative sources of energy like solar, wind, hydro, geothermal, and tidal are still in early development. These sources cannot compete with established sources like coal and oil, at their present stage of evolution.

Coal and oil, have ruled the world for a long time. We have dug, burrowed, and mined our planet to our heart's content. The impact is nothing short of catastrophic. It is not only global warming we are concerned about, but also the non-sustainable form of energy source on which our global economy depends.

The data below shows the dominance of fossil fuel over other sources in the world.

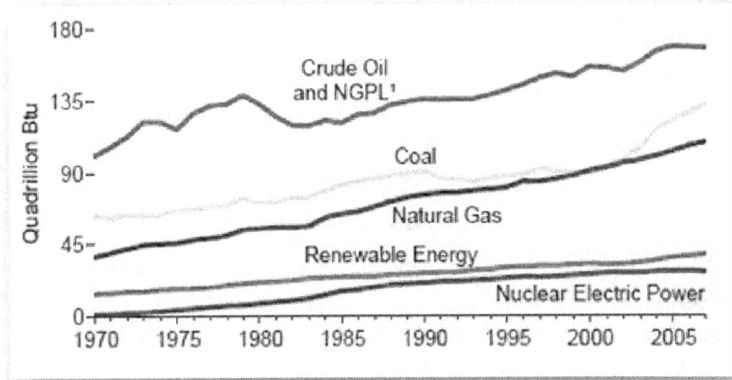

Fig. 12: World primary energy production by source

To find a solution to this energy crisis, let us see how our energy industry evolved all these decades. Initially energy industry started with very primitive thermal power plants using coal from small mines. Even the oilrigs that extracted oil were very small. As more factories/cities emerged, the market size for energy increased. This market demand put a lot of pressure on the quantity of energy generated. The demand for more power gave rise to the evolution criterion *throughput* or *scalability*.

To satisfy this criterion, the energy industry responded by building larger thermal power plants and more oil refineries. The shipping and transportation industry expanded as well to transport coal and oil. Each thermal power plant came with a huge investment in coalmines. Oil rigs and refineries expanded. The profits were also equally high.

The other factor that is of greatest concern is the participation of developing and underdeveloped countries in the global economy. These countries have begun to consume considerable energy, especially oil and coal. We need under-developed countries to develop faster to support the growing economy.

As the global economy expanded, demand for energy increased so much the energy industry was unable to meet the demand quickly enough. It takes years and decades to build thermal power plants and refineries. The sudden increase in demand and insufficient supply caused the cost to rise. We have hit many energy inflicted recessions in the past.

As futurists, should think of addressing throughput or scalability issues. We can use automation to ease the scalability problem. We have two possible options: We can automate existing processes in refineries and thermal plants or use it in future energy sources like solar. If we plan on using automation in existing coal mines, thermal power plants, oil rigs, and oil refineries, we can eliminate the human element and replace them with efficient machines and robots that will not only do a faster, more efficient job but could be replicated to build more power plants simultaneously. This will improve the supply and the cost will decrease substantially. However, these processes are highly complex and automation will be equally expensive and also these sources (oil and coal) themselves are being depleted and causing ecological problems.

Back when we first struck oil, we knew oil would one day be depleted. Still we celebrated it and built mega industries around this non-sustainable source. Entrepreneurs may or may not take us on the right path; they are there to compete and succeed in business. It is the responsibility of governments to make the appropriate policies and guide economic growth in the right direction. This is where all governments around the world failed. We should not do the same mistake again, going forward the more prudent path is not only to switch to solar energy but to automate the process of building the solar power plant.

The sun has so much energy that merely capturing an infinitely small fraction of it will make us energy sufficient. Tapping into such a mega source is the wisest decision we could ever make.

Automation in Energy Production

The usage of robots in any industry will not happen unless the ROI (return on investment) is profitable. Robots are still very expensive and automation of production units requires both time and money. This is the reason many businesses, including energy producing companies, still use manual labor and struggle to increase their throughput.

We need to exploit the amazing feature of automation which is its ability to replicate. This is exactly what we need in the energy industry today. We will be able to produce as much energy as needed and as quickly as possible by replication and not by building each power plant from scratch. Several solar power plants have been built in the world from scratch, each one a massive project. Transferring these skills to other humans via training is quite a task.

If we build power plants with robots, they however, will keep all the programmed skills and take the stress, anxiety, and human error out of training and teaching workers to do the repeatable skills a robot could do both precisely and continuously.

If we use robots in every individual processes, like manufacturing solar cells, assembling solar panels, the transportation of these panels to a power plant site, and finally to the assembly of a solar power plant on site, then we will end up with a fully automated power plant building industry, which can sustain itself and could be replicated when needed. Solar power plants will now become the end product of this massive industry. In essence, automation transforms mega projects into products that can be mass-produced.

These robots, once programmed to do specific tasks, will keep doing them tirelessly until the power plant assembly was complete. Initially, of course, the cost will be very high, as highly skilled labor to plan, build, and program these robots is needed. We might even need substantial energy from fossil fuels to build such a plant. After the completion of the first plant, however, the entire army of robots will build another solar power plant, an exact replica of the first, in another designated location, by themselves.

There will be no further need to start the process from scratch or enlist human talent to design or redesign new blueprints or drawings. In fact, no skill transfer will be needed, period. All the information fed into these robots during the first power plant would be reused in the second plant, except perhaps for some new information about the topography of the second site. We could replicate the entire army of robots to build several power plants in parallel.

Imagine an army of robots producing several solar power plants simultaneously without much human physical involvement. Imagine an industry that could produce such an army of robots continuously. Such automated factories will resemble the Japanese *lights-out* factories. Within a short time we would end up with thousands of global power plants all built and managed by robots.

Although this process is fully automated, these automated industries need enormous skilled human labor to monitor the robots and to redesign or reprogram them for better efficiency. These robots go through continuous improvement as humans make them better day after day. This is how the industry will technologically evolve into the future and human skill level also increases that supports higher standard of living.

The ultimate benefit of automation is its quick replication capability to produce power upon demand, addressing the throughput or scalability concerns that we have today.

What will be the cost of electricity produced by this industry? It will be the cheapest possible; the cheapest, in fact, we have ever paid for such a plentiful and reliable energy supply. As technology behind solar cells becomes even more sophisticated, solar energy will become the most affordable on the planet, taking the global economy to new heights.

By the time we reach the end of the Age of Automation, energy will literally be free. Like ants building ant-nests or bees building honeycombs without any human effort, solar plants grow in number without any human intervention.

Once the robot-manufacturing industry establishes itself, robots will be very affordable and other businesses will naturally start adapting automation methodology. The biggest impact of this will be on raw materials. Currently, as mining companies use manual labor and expensive energy, the raw materials are very expensive. With cheap energy, the raw materials will become very affordable. When the miners adapt the automation method, the raw materials will become even more affordable, making our end products inexpensive.

Compare the above automation approach to the manual approach currently employed. Expensive human labor is involved in producing solar cells, assembly into solar panels, and designing and building solar power plants. This labor-intensive process not only makes the production slow and inefficient, but extremely expensive and unable to compete with oil or coal.

A great promising technology of the future is 3D printing. The current 3D printing technology could evolve to 3d-print power plants in the future. Another promising technology in the energy sector is hydrogen fusion, which is not only green but potentially very affordable. This is still in the research stage and there is a long way to go before seeing practical results. Nanotechnology is also promising, which will be able to absorb, store, and deliver energy at a very affordable price.

These and many other upcoming technologies could successfully replace the traditional fossil fuels. It is hard to say which technology will succeed, but they all will adapt automation to solve the scalability problem.

As futurists, it's very interesting to envision how cheap energy will impact our lives. Energy is needed pretty much everywhere. If it becomes free, all products and services will be highly affordable or may even be free, leading us to wonder: do we need to work at all in the future? Well, if money is the motive behind hard work, rich people would have stopped working a long time ago. Surprisingly they work harder than others. Bottom line is, the world as we know it will not end with cheap energy; it will only grow stronger.

CHAPTER 11

THE FUTURE OF HOMES

What if I told you a house that once took months to build could be constructed in days or even hours? Our next evolutionary era, the Age of Automation, will find us building mega cities in months, maybe even weeks, as opposed to decades or centuries as in the past.

Though competition is holding construction costs down, technology is gradually pushing the cost upward. Skills like carpentry, plumbing, and electrical work are highly paid because of the technology involved in handling the sophisticated tools, materials, and appliances. Once we move out of manual labor-intensive and skill-dependent industries, the cost of construction and maintenance will plummet.

One such technology helping us cross over is 3D printing. Fortunately, the 3D printers are evolving much faster than expected into house building devices. In china, there are already such machines that can build homes in one day. Experiments are also going on to build multi-story homes with machines.

Using the embedded digital sensors, the health of the house structure will be monitored by computers. With the advanced technology readily present in the Age of Automation, every electric device in the house will have its own energy source, making external power unnecessary. Vacuum-drying sewage eliminates the need for disposal via pipes. It also controls many diseases from spreading as the sewage system is fully eliminated. Even the drinking water is recycled eliminating the dependency on our ecosystem to clean the water we use.

In the future all fixtures including those to heating, air conditioning, washing machines, and even refrigerators are just screw-on items placed on the outside of the house. This will make it easy for robots to replace them as needed.

As technology improves, we will change wall color digitally to satisfy the desire for choices and walls will have digital displays and TVs. Crawling, mouse-sized robocleaners will keep the house clean. If there's a water leakage or fire in the house, sensors will identify it instantly and attend to it before the homeowner is aware there was a problem.

These futuristic, green technologies will emerge as we progress into the future, where living an eco-free life will mean no help from Mother Nature to clean up after us.

These 3D-printed houses will be stronger than their man-made wooden counterparts. They will also be demolished and rebuilt quickly so losing a home to a hurricane, earthquake, or other disaster will not be the catastrophic financial loss it is today. In fact, with advanced technology you would have moved to another home before a hurricane hits!

In the Age of Automation, by using soil as the main raw material, we could move away from dependency on forests. As the technology matures, competition will produce stronger and cheaper houses.

Automated Kitchen in Homes

As people are getting busier with work, daily cooking at home has become more of a hassle. Our dependency on canned and frozen food will continue to increase. However, human desire for freshly cooked food will never diminish. We will see a substantial growth in technologies and businesses that help us save time-to-meal. There are already a host of startups catering to this new trend. In the US, Munchery lets customers order food using a mobile app and meals are delivered within about 30 minutes.

Another early similar food delivery startup is Blue Apron, which delivers a recipe along with ingredients for the meal so the customers don't have to go shopping. Other early entrants to this market include ChowNow, DoorDash, and Hello Fresh. UK-based Whisk has developed an app that builds a shopping list for any recipe you pick online. These kinds of apps are increasing in number and are designed to learn a great deal about their customers' tastes. The intelligence built into these apps is helping people make healthier choices, save time, and have fun cooking new dishes they haven't tried before.

In 2014, food-tech companies raised more than $1 billion in the US alone. They are also expanding very fast in developing countries like China and India.

This trend to satisfy human desire for more variety and less time-to-meal is taking us to the next level which is the cookless kitchen. Moley Robotics has developed an automated kitchen with robotic arms that can mimic a human chef's actions to cook the meal. Although this is not a full-fledged cookless kitchen, it is a step towards it. Another company called June has created an oven that recognizes the food placed in it and then cooks it perfectly.

As technology improves, artificial intelligence to cook food will develop very quickly as there a huge demand. In the future homes, the entire kitchen will become an appliance.

Thousands of worldwide recipes can be stored in this kitchen machine and thousands more could be downloaded from the Internet when needed. Just choose the recipe and the robocook will prepare chef-quality food. Meals will be categorized much like a song play list on your computer: by weekday, holiday, time of day, time of year, food groups, or even family members' favorite recipes.

If you select the menu ahead for an entire week, the machine will order any additional ingredients needed in advance. If you plan a weekend party, the machine will be ready to dish out the food. In addition, depending on your health needs, the smart kitchen software will suggest certain recipes or avoid specific ingredients.

Delivery robots will load groceries and vegetables into your kitchen machine from the outside. As a homeowner, you will have no regular involvement in maintaining the grocery inventory at all. Everything is taken care of by the machines, saving you hours of time and producing a great meal every time.

Meet Your Home Manager

As we become comfortable with technology, managing our busy lifestyles with time-saving features becomes exceedingly important, making the need for an intelligent home inevitable.

A Future home is a collection of "IoT." Internet of Things (IoT), refers to the ever-growing network of smart objects—smart phones, smart appliances, smart watches, smart televisions, smart cars, smart street lights—that have Internet connectivity so that the devices can communicate with one another. The evolution of the smart phones is driving the development of the IoT.

A *thing*, in the Internet of Things, can also be a person with a heart monitor implant, a farm animal with a biochip transponder, or any object that can be assigned an IP address to transfer data over a network.

The IoT is intended to give us more control over our lives. A smart car can be tracked. An owner can start their car remotely and turn on the heat so it's cozy warm on a winter morning by the time the owner gets in to drive to work. A smart or intelligent home can be remotely monitored to adjust the thermostat, close or open blinds, turn lights on or off, or access cameras so the homeowner can check in on kids or pets.

Conceptually, every object on the planet could be assigned an IP address. New lines of smart appliances currently for sale have joined the Internet of things. Consumers can now communicate with stoves, refrigerators, and washers and dryers. More importantly, the appliances can talk back. Smart refrigerators can inform a homeowner when they are low on milk or need eggs. Some smart refrigerators will also suggest that night's dinner based on its contents. In the future all home appliances will become interactive, linked electronically to a home-management software or a home manager, designed to understand our habits and maintain our homes accordingly by controlling all the smart devices at its disposal.

In the event of a fire, for instance, the home manager will identify the exact spot via temperature sensors and video cameras and will put out the fire with built-in sprinklers or a sprinkler-attached drone designed for that purpose, long before the fire engine arrives. The manager will report malfunctioning appliances to the dealers who will replace them robotically, and will even vacuum the floor by deploying crawling micro-robots. The list of services is nearly endless.

The demand for smart homes will be so great, new technologies will emerge as companies compete for their market share. As energy becomes cheaper, these technologies will grow exponentially. As futurists we should be able to see a better world with smart homes. A world that is capable of protecting us during hurricanes, tornadoes, and other natural calamities. A home should truly become a comfy nest.

CHAPTER 12

THE FUTURE OF REAL ESTATE

Homes, although built primarily to live in, have become investment properties as well. Many people bought homes with the idea of building equity while living in it. It is a dream not just to own a home but also to grow that investment. As prices rose rapidly during 2001-2006 period, many homeowners became serious investors and started buying second and third homes as investments. There is nothing wrong with such investments as long as folks can afford them. However, this caused more demand in the housing market and prices started climbing because of a fictional perception these homes were actually worth the inflated amount.

In a booming market, even if the homeowner is unable to pay, the home value will have appreciated and the banks will recoup their money by selling those homes in the event borrowers default on the loan. In a seller's market, even if the homeowners are a credit risk, the houses themselves are not. This is one of the reasons for promoting easy credit.

Below is a graph showing the United States home price variations (bubble) from 1988—2008.

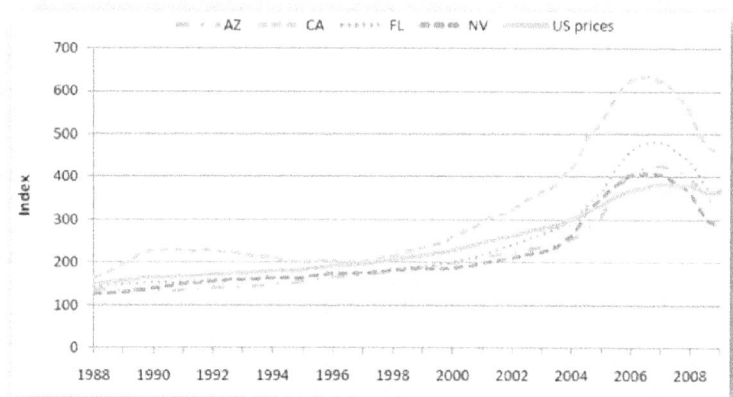

Fig. 13: House price appreciation by selected states

This speculative housing market is no different from the stock market. When the fictitious price of houses deviates too much from the real prices, people feel the pinch and slow down buying. Fed in the meantime also increased the interest rates to control inflation making it harder to pay the mortgage premiums.

When the market becomes slow, real estate investors start selling their homes to recoup their investment creating a buyer's market. As a natural turn of events, house prices decline.

Though the housing industry has all the characteristics of a stock market, houses are not sold as frequently as stocks. While stocks are traded every day, people typically sell their homes every three to five years. This is the reason why a real estate crash will not happen in one day. Real estate market crashes are slow and prolonged for years.

When house prices go down, banks worry about the loans they have made to risky buyers. When such houses go into foreclosure, banks lose substantial amounts of money. The amount lost depends on the number of risky loans they may have awarded and the size of the bubble was. Some investors take equity loans on their homes and invest in stock market or sell stocks to buy homes, depending on the market conditions. Because of this interdependency, the housing bubble and stock bubble affect each other. A housing bubble combined with a stock bubble could cause a serious economic slowdown.

Obviously a less risky loan is the answer. The banks are not assessing how risky their own investments have become until the customers begin to fall behind their payments or homes begin to foreclose. For a safe and healthy housing market, it all comes down to how frequently the banks reevaluate the value of their awarded mortgages.

Now imagine a bank giving out loans to risky customers with very limited background check on them. When the economy slows, it should not surprise anyone if these banks go bankrupt. This is exactly what happened during the real estate crash of 2006.

Now the question is why our banks are not monitoring or managing the utility values of their investment.

Automation in Mortgage

Human involvement in making a decision on a loan makes it an inefficient process, as well as increasing chances of error, either intentional or otherwise. What we need today is a consolidated financial database to track down the financial status of every borrower, be it a person or a company, to identify the risk before awarding a loan. We need to develop a system that monitor factors related to the borrower such as net savings, future earnings potential, the stability of the industry employing the borrower, and so on. The more accurate the information it gathers, the more secure the mortgage would be.

With automation, most bank loans will be less risky as the software will instantly evaluate the applicant's financial ability without subterfuge or human error.

Even after the loans are awarded, the system should continue to evaluate the mortgages' utility value based on the changing global economy and the banks can figure out how risky their loans are, well before the customers fall behind their payments. When the economy fluctuates, the system should identify the customers that belong to the risky category. How well they identify the future risky customers depends on how extensive is the data collected from the global economy and at what speed.

More sophisticated the mortgage system become, they will be intelligent enough to identify the bubbles forming in the real estate well in advance. The banks can put more restrictions like higher down payments in those areas where bubble is forming.

With software-controlled regulation in place, many activities like sub-prime mortgages will become obsolete. With this kind of closely monitored money lending, the housing market will grow gradually and will encounter gentler price adjustments than a bubble followed by an inevitable crash.

Compare this futuristic software regulated real estate market to the current one. When prices begin to climb, all investors jump in to invest causing huge bubbles. Not only buyers suffer from losses when the bubble burst, banks are also affected hurting the economy.

It has always been a human dream to acquire more wealth. This desire has been and will continue to be the backbone of our strong economy. However, acquiring wealth itself should not hurt the economy.

In today's booming global market, customers will invest locally as well as overseas where they feel there is growth potential. Don't be surprised if the owner of your neighbor's home lives in some other country, collecting rent electronically. People will buy and sell homes around the world, a trend that will predominate in the near future, as more and more underdeveloped countries prosper due to outsourcing and entry into the global market.

To monitor international investors we need global financial institutions to share data and work together to develop and maintain a robust lending system to filter out risky investors globally. Although this is the only approach to avoid future global real estate crashes, this needs lot of coordination at international level which is normally a slow process and let's hope this will happen at relatively faster pace.

Future of Real Estate

There are four disruptive technologies that will decide the fate of real estate in the future. The first is autonomous cars. Crowded cities around the world are a nightmare for daily commuters. The self-driving cars will ease traffic congestions considerably, enabling people to relax while commuting. Commute distance will no longer be as grave a concern. This will prompt city dwellers to move further away from crowded urban areas to where they can improve their standard and quality of living.

The second disruptive technology is 3D printing that will build affordable houses in matter of days. When 3D printers become commercially viable, home prices drop to the actual construction value because supply will quickly meet the demand. Future homes will not only be highly affordable, but will be recycled often and climb the evolutionary ladder. In the future it would not be uncommon to see people gutting and rebuilding their homes several times in their lifetime. We could extrapolate the same idea to building new cities as repairing old cities would become expensive compare to building new ones. With 3D printing technology, we can quickly rebuild cities when old ones are destroyed by natural catastrophes.

The third disruptive technology is high-speed transport. If you can travel 100 miles in 10 minutes do you really care about your daily commute? This technology will enable us to move further away from crowded cities, but again, not too far from hospitals and healthcare facilities. The concepts of hyperloop and gravity powered transportation will have phenomenal effects on real estate when they become realities.

The fourth and most disruptive technology is automated healthcare technology where machines will take care of our health in the comfort of your home. With this technology, which is a true game changer, a revolutionary approach to human dwelling, homebuyers will aspire for more comforts and luxuries in their homes than proximity to cities or towns. We will begin to see new towns and cities built by machines all over the world in remote locations. By this stage, humans will have learned to shield themselves from natural catastrophes like hurricanes and tornadoes. Not only would have the forecast methodology improved but also the rate at which humans recover from them.

In the Age of Automation, living in crowded cities would be a thing of the past. Every individual will be technologically capable of living anywhere on earth, following their hearts desire. As humans are social beings by nature, cities may not go away entirely, but the quest to live independently will continue to affect the real estate market.

The other important factor that affects real estate in the future is Global warming. As ocean level rises many coastal cities will become unlivable. They will have to be relocated and some, rebuilt entirely. Similarly, flood prone regions that are at the mercy of overflowing rivers and lakes will need to be relocated as well. As nature's fury is unleashed with wild hurricanes and tornadoes, the way we build homes will be inherently different in the future.

New technologies are currently only being developed to satisfy the human desire to live luxuriously, with modern amenities and away from crowded cities. However, in the future, the two distinct driving forces – our desire to live luxuriously and our struggle to survive from global warming catastrophes – will affect the rate and direction of technological innovation in home construction, and in turn, shaping the real estate market.

CHAPTER 13

THE FUTURE OF TRANSPORTATION

The world is waiting for self-driving cars. Taxis soon will disappear from the street altogether. In fact, very soon, we may no longer buy cars anymore. Instead, we will just hop into a self-driving car on the street and punch in our destination.

Though most drivers are law-abiding citizens, human error is inevitable when people are distracted or stressed; accidents happen and many times, are fatal. Considering the stress we are all going through, frankly, no one wants to drive, especially to work.

Before looking at the future of cars, let's see how our cars evolved all these years. The primary criterion for the invention of a car is to transport. Automobile companies have designed and built a variety of vehicles to satisfy this desire.

Then came evolution criteria: safety, comfort, and speed. Many safety devices have been invented, giving rise to such features as air bags, ABS, and so on. Many comfort features evolved like air conditioning, power windows, power steering, cruising and so on. The demand for high-speed cars was always there, we could only achieve higher speed within the vehicle's stability level.

Now with better technology and better roads, although we are able to achieve relatively higher speeds, human driving is simply not able to keep up with the demand. Simple mistakes lead to deadly accidents at higher speeds. Humans are not able to cope with this intense pressure while driving. No matter what safety features we have, the speed works against it. Even with the speeding ticket system, our traffic has not slowed and accidents are on the rise. Human error is the culprit behind most road accidents.

The solution to this problem is to gradually reduce and then eliminate human involvement in driving by introducing intelligence into our cars. This is already happening. We are already seeing many features like collision avoidance, blind spot detectors, lane change detectors etc.

Many car companies including Mercedes, BMW, and Tesla have already or soon to release semi-autonomous versions of their models. By 2020, the experts are predicting that nearly 10 million cars will have at least one semi-autonomous feature and some would be fully autonomous..

The barriers to self-driving cars remain significant. Costs need to come down and regulations need to be clarified around certain self-driving car features before the vehicles fully take off. As demand for self-driving cars is very high, these milestones may be met earlier than later.

Lifestyle Changes with Robot-Cars

Self-driving cars will dramatically change our way of living. As no drivers will be needed, driverless taxis will become so affordable people will use them for daily transportation rather than buying their own cars. This will be a big relief from car ownership and maintenance chores! Service providers will compete with each other to provide state of the art cars on the roads for people to hire.

As traffic will be completely controlled by computers, it will be highly predictable. They could even predict traffic before it happened, as traffic monitoring devices will have the history of all car movements. Many employees go to the office and back home in a very predictable fashion. When people punch in their destination, the Car-computers will reroute traffic to alternate less busy routes.

Now, with these robot-cars, people may not need parking spots any more. As there is no car ownership, there is no need to park cars. However, for people carrying their personal belongings may need some temporary parking. It means that the parking system will not completely go away, but will substantially reduce in size. Our garages will eventually become storage rooms for robotic-goods delivery system.

Self-driving cars are a boon to elderly and sick people. They are great for kids and young adults who are not eligible for a driver license yet. It would definitely be loved by travelers going on sightseeing tours. As they don't need to drive, they can just sit back and enjoy the tour. Shopping, goods delivery would take new shape. You can get dropped off at the door of the shopping mall and picked up at the door again. If you have more things to carry, no worries, put them in another self-driving cargo car send it.

In the future we may not need to learn driving. May be the skills of manual driving will become obsolete. However, all this will not happen overnight. It will take decades for this change. But we will begin to see this technology creeping into our lives very soon.

Robotic Road Repair

When self-driving cars are used in road repair, the results would be dramatic. These automated machines can repair the roads as they roll on the street, without any human intervention. Is this technology disruptive enough to cause considerable job losses? Well, in reality, a considerable number of people are needed to design and assemble these machines. This opens up another fast growing industry. They hire people who have the technical knowledge of road repair. The current employees who have the knowledge are an asset to those companies to help build the machines to do all the manual work they themselves used to do.

You may argue these automated industries may not absorb all the laid-off workers in that industry. We may see machines replacing humans, but we fail to see the full impact of machine pace. When we automate any job, not only does its cost go down, the pace of production increases exponentially. Just like automating a bottling process that produces millions of bottles a day or automating a textile company producing thousands of dresses each day. The same logic is applied to road repair as well.

Once the machines are built, they are easy to replicate. They can repair thousands of roads at the same time, not one by one like is done today by blocking roads off for weeks or months. The machines not only do this job continuously without a break, but they also can be scheduled at will. You can put them on the job or take them off at your discretion. You can pull them off during peak hours and make them work at night. Once the automatic machines come to market, we will see many roads being built or repaired simultaneously on a regular basis instead of once every several years. Just imagine, a thousand machines working all night and seeing a brand new road in the morning. A pothole you spotted on one day is not seen the next day. An automatic repair machine that already scanned the road has repaired it.

As the machines are easy to replicate, there will be thousands of machines at work while we sleep. Compare that to our current road repair method — hundreds of workers struggling with old machines and at the mercy of high-speed traffic. Even traffic suffers with congestion during the repair season. Roads are blocked for months or even years.

Dynamic Lanes

Many times while driving on highways, we see the other side either clogged or with relatively less traffic. This is because the lanes are mostly fixed and are not dynamically usable. Some municipalities and states have addressed the issue by introducing express lanes, which are often empty or inefficiently used and by flex-lanes, which change direction depending on the time of day. But this will change on a national scale when smart cars drive our roads. All lanes will become interchangeable dynamically and give more room to congested traffic. The same will be true for surface streets in cities. Cars from both sides will seamlessly merge and move on. Pedestrians will cross the roads without fear as cars will automatically stop for them. The best news is that there will be no drivers to get upset! A sight that we have never seen before.

Cars and Drones

When self-driving cars flood the market, they will offer other services as well such as delivery services. All items our home robot-manager orders will be delivered to our homes via delivery robocars.

The drone delivery technology that created quite a stir in the news media had a setback in the US when the Federal Aviation Administration proposed rules preventing drones from flying outside of the sightline of the operator. However, Canada, the U.K. and Denmark are already using drones commercially.

Self-driving vehicles and drones could make a good team supporting delivery services. Drones could load the self-driving vehicle, travel with them and then when cars reach their destination, the drones could wake up and deliver the goods. These two technologies could complement each other to achieve an efficient delivery system.

Self-driving vehicles could also team up with the 3D printers that build houses to deliver the goods to the construction locations. We may begin to see trucks with robotic arms to load and deliver the goods.

Self-driving cars could also team up with trains and buses, and drop you off to train stations or bus stations. Cars can even be programmed to come to your house every day at pre-determined time and pick you up and let you know if the train or bus is delayed so you don't have to wait at the stations. The train and buses will know exactly how many passengers are coming. All this public transportation will gradually become intelligent giving exceptionally good service.

The self-driving cars can also team up with airlines to drop you off at the airport at the right departure time and pick you up at your destination. Vacation travel will become much less stressful, especially for the elderly and families with kids.

With self-driving cars will there be less traffic congestion? Surely this disruptive technology will ease the traffic. First of all the trucks that are self-driving could be operated at night. This will substantially reduce the traffic. Many other delivery trucks will probably follow similar arrangement. During peak hours, smart cars could take alternate routes further reducing the traffic. Gradually, traffic will monitor and manage itself. It can't get any better.

CHAPTER 14

THE FUTURE OF THE EDUCATION SYSTEM

Though our world is growing at an unprecedented pace, higher education is still a dream for many. Colleges are so expensive many families cannot afford it. Data from the National Center for Public Policy and Higher Education shows the increasing cost of education. The increase in the price of college has outstripped price increases in other sectors of the economy.

In October 2013, only 65.9 percent of high school graduates were enrolled in colleges or universities, according to data released by the Bureau of Labor Statistics of the United States Department of Labor. Financial difficulty is found to be the primary reason for this.

To compete, colleges are constantly adding infrastructure, making education more expensive. Be it a state-of-the-art laboratory or sports facility, the cost raises as technology improves. Rising teacher salaries have also added to the cost.

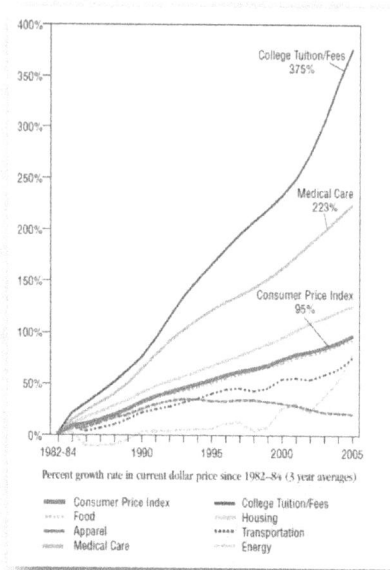

Fig. 14: Increase in the price of college and other sectors of the economy.
Source: The National Center for Public Policy and Higher Education

Looked at from an evolutionary angle, we know any product or service will become affordable if it goes global. Our colleges are no different. The good news is there is no need to globalize our colleges from scratch. Many colleges are already online, offering courses that do not require students to attend a campus. Many students and employees are taking online education for various reasons from enhancing their career path to simply gaining knowledge.

The advantages of online classes are affordability, flexibility and scalability. Today many colleges and universities are depending on fees from fulltime students and some financial assistance from governments. When they open their doors to the global market the revenue will be substantial.

Competition in Teaching

Excellent teachers are rare. Students will greatly benefit from great teachers' lectures made available online. This applies to both on-campus and online students. As we have seen, at colleges, students flock to popular classes while others are scarcely attended. By putting the classes taught by good teachers online, colleges benefit all students globally. Their very online presence will make a difference to the education system. Students may access online classes to supplement their regular courses for a nominal fee. This will lead to an unprecedented demand for good teachers. Becoming an online professor will soon be a very high-paying profession and motivate teachers to be better.

With the potential online revenue at stake, colleges will compete to offer the best online education. Students will learn much faster by attending these competing online colleges around the world. While students look for the best online classes, colleges look for the best teachers. Imagine the effect of such high quality and low cost education on the world's technological evolution.

By globalizing the educational system, colleges increase the market size and earn more revenue, bringing down the cost of education and making global education possible.

The Future of Education

Concerns surrounding the lack of a physical presence in an online learning environment have led researchers to investigate the concept of virtual presence. Many colleges are currently using tools like discussion forums, chat rooms, email communication, and so on. However, the gap is hard to fill with these tools.

Many studies have suggested that learning occurs through the interaction of social presence, cognitive presence, and teaching presence within a community of teachers and students. The inability to participate in group discussions, lack of face-to-face communication, and absence of practical training are the biggest challenges that online colleges are facing today.

One of the upcoming technologies that would help online colleges is virtual reality (VR), the same technology already popular among video gamers. There is already an intense competition to grab this new market including Face Book's Oculus Rift, Google's Cardboard and Jump, Sony's PlayStation VR, Microsoft HoloLens and HTC Vive. Many others are on their way. As technology evolve, VR technology via specialized headsets will enable online schools and colleges to provide practical, immersive training to students through a video game-like experience. Many companies are already using similar devices to provide virtual walking tours of colleges and universities.

In manufacturing VR is also becoming a powerful educational tool to develop products and train employees. Automobile companies like General Motors and Ford already use VR systems to communicate between designers, vendors, and assemblers. Designers work in a virtual 3-D medium, get feedback from other departments, and make adjustments in real time.

The VR market is expected to reach more than 25 million users by 2018.

In the future, after basic schooling students will be trained in specific areas of expertise to support global companies. *Customized curriculum* will become the future of education, which is not only more efficient for students but also greatly benefits global companies as the skill demands are immediately met. In the current education system, students are forced to study subjects that often have no relevance to their areas of expertise or their desired training so it can take decades for them to develop into skillful employees. This is a system on the verge of a paradigm shift. As speed of education becomes paramount, the driving force of the global education system will be customized, rapid training.

When our currently rigid education system becomes flexible enough to cater to the needs of the market, it's inevitable that global recruiting companies will provide input to customize curriculums to suit their business needs. So if a company is planning to hire 10,000 employees, it makes sense to prepare them to learn those specific skills. The need for accelerated learning and customized curriculum will bring colleges and employers much closer.

Now the question becomes, what will be the future of our education system if the curriculum is controlled by the recruiting companies. Will colleges lose their freedom of teaching traditional curriculum? What if someone wants to learn music or sculpting for an independent career or just as a hobby? Would online education take that market? There are already such systems and apps that cater to those needs. There also many individuals and small institutions offering many such courses online. The only piece missing is the certification and quality recognition. This is where traditional colleges will continue to play their role. Colleges may eventually be reduced to testing and certifying agencies. As the current rigid system loosens up, we will see a new market emerging making the education system truly global.

CHAPTER 15

FUTURE OF OUTSOURCING

Outsourcing has become a most controversial issue and highly debated subject in the world today. In this chapter, we will look at outsourcing from an evolutionary angle and analyze if this is really taking us in the right direction.

Two major categories of jobs that are getting outsourced are blue-collar and white-collar jobs. The outsourced blue-collar jobs include all manufacturing jobs. The outsourced white-collar jobs include all skilled jobs related to Information Technology, telecommunication, biotechnology research, financial analysts, architectural designers and drafters, telemarketers, accountants, medical billing, medical transcriptions, claims adjusters, home loan processors and so on. The list of jobs keeps changing as the factors responsible for outsourcing changes.

There are three factors responsible for outsourcing: labor cost, skill availability, and the speed of information technology. While low labor cost encourages blue-collar job outsourcing, low skill availability promotes white-collar job outsourcing. If a country offers highly skilled, low cost labor and information technology is sophisticated enough for fast communication, then white collar job outsourcing happens seamlessly.

Let us discuss both blue and while-collar categories and examine their effect on our economy.

Effect of Blue-Collar Outsourcing

American companies could have chosen automation instead of outsourcing to reduce the cost of products and services. Unfortunately, many companies chose outsourcing option as the initial cost of automation is very high. If they had selected automation option, not only it would have created more skilled jobs here but also increased the pace of technological growth and at the same time increase the standard of living. The same automation technology would have spread around the world making the global economy much stronger. However, that did not happen. Jobs got outsourced to countries where labor cost is low.

The current blue-collar outsourcing has both positive and negative effects on our economy. Let us look at an example and see how it effects on our economy. Say a private company manufacturing shoes with local workers sells a shoe for $50. If that company outsourced its jobs to another country, the cost of a shoe might be cut to $20 or less.

The negative impact is partial or complete layoffs depending on the extent of the outsourcing. The price reduction will put pressure on other shoe manufacturers to outsource their jobs as well, increasing the number of layoffs. The employees who lost their jobs will need to be retrained to be able to join the work force, either in the same company at a different level or in a different company altogether. They will have an opportunity to change their career, if they were able to adapt. If the laid off employees were older, then the transition will be more difficult.

The positive impact of outsourcing is the increased market size because of the price reduction. Even poor people will be able to afford shoes at this price level. More sales mean more economic activity.

This impact is from just one product. As more and more low cost products come into the market, the market size will expand, and the international shipping industry will become busy and the domestic transportation system will improve. These expanded industries will hire more workers. Increased sales will bring more shopping malls and expand existing malls, offering construction and maintenance jobs to many workers. The expanded shopping complexes will hire sales employees and support other services like food courts, creating a ripple effect in other business sectors.

Increased sales will affect the credit card industry as well, with more transactions and as a result, more revenue. This industry will hire more people as their operations expanded. Banks will open more ATM machines, as the need for cash to make purchases increases. Banks will hire employees for various operations.

Every other industry will benefit to some extent by a reduction in prices and increased sales. Price reductions help poor people, too, as even they begin to shop as a result of their improved standard of living.

Lower prices substantially increase the living standard of people. For example, shoppers who used to buy two pair of shoes for $100 will now buy five pairs or more on the same $100 budget. For arguments sake, instead of buying one hundred products for $1,000, shoppers will be able to buy three hundred to four hundred products for the same amount. This increases their standard of living substantially. Again, it is not just products but also an entire ripple effect that will branch out across multiple industries and income streams. In fact, this is equivalent to seeing three hundred to four hundred shops open in place of one hundred shops, leading to more activity in all other services discussed above.

Today, outsourcing has become so common new companies consider outsourcing from day one. Part of their organizational blueprint includes the costs—and savings—of outsourcing key positions, particularly blue-collar labor. Businesses may not even open their fabrication units in developed countries, if they do, they may not get bank loans as they can't beat the competition. Outsourcing has become a survival strategy.

Can We Stop Outsourcing?

Many experts strongly advocate that the government must make it difficult for employers to outsource. There is also a strong notion among many economists that if we can somehow stop outsourcing, the local companies will create more jobs here in America. Would it? Outsourcing has happened for decades; it is not something new. In fact, from early 2000 all the way to 2007, outsourcing was at its peak, resulting in substantial job losses. Even so, the unemployment rate decreased during this period because most people who lost their jobs were able to find local service jobs, which had been created due to higher economic activity.

Good news is, nearly 70 percent of the jobs are hard to outsource and they fall under local service jobs category. Only 30 percent of the jobs have the potential to get outsourced and those are blue-collar jobs. According to the Department of Labor's Congressional Budget Office, the United States economy created about seven and a half million net new jobs between early 2004 and the end of 2007.

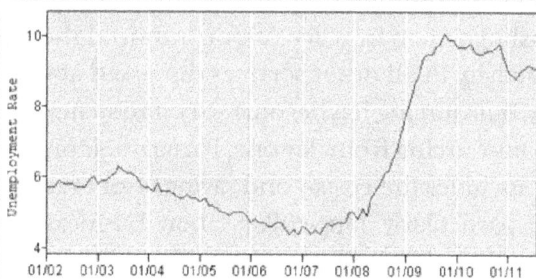

Source: U.S. Department of Labor
Bureau of Labor Statistics

Fig. 15a: Labor force statistics, Jan 2002 - Jan 2011

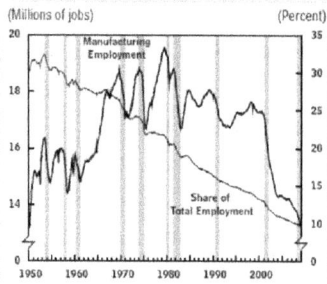

Source: Congressional Budget Office based on data from Department of Labor, Bureau of Labor Statistics.

Note: Data are quarterly and plotted through the third quarter of 2008. The vertical bars indicate periods of recession, as determined by the National Bureau of Economic Research.

Fig. 15b: Rise and fall of manufacturing jobs in United States

The above two graphs show the reduction in the unemployment rate during the intense outsourcing period between 2004 and 2007. If the same trend had continued, we would have moved much closer to the age of automation and most of the jobs would have come back to America, not in terms of low paying manual jobs but in terms of high paying automation jobs. Unfortunately, the recession slowed down the job creation and unemployment soured. However, the outsourcing did not stop. It continued to move jobs out of America.

If the government made some policy changes, including tax incentives, some blue-collar jobs may come back. How many of them will last is unknown, as is what will happen to such jobs when employers find other countries where labor is much cheaper than China and India.

Is there a way out of this crisis? Automation is the right strategy to bring jobs back into the country. We need to promote automation in all industries. Fortunately 3-D printing and robotics are evolving and are taking us in the right direction. Automation in energy production will give us the most desired results, but can automation happen on a mega scale with limited availability of skilled force?

This means our education system and its curriculum must evolve to retrain unskilled workers so the automation methodology can be used.

While adapting automation, technology behind automation can also be sold to other countries to recoup the investment. Massive job creation, cheap green energy, and faster power generation will pave the path for faster economic growth.

No matter how we look at the economy, the pace of technological evolution is fully dependent on how fast our current rigid education system would evolve. As long as automation is in focus, we will be on the right track.

Automation in Other Countries

Even in countries like China and India and other emerging markets, as the labor costs increase, automation looks more lucrative than traditional manual labor. As demand increases, the speed of manufacture and delivery of service will become the driving evolutionary criterion. Manual production can't satisfy these needs. Inevitably, all these countries will be treading a path of no return. Once they enter the automation mode, they will embark on a new journey into the age of automation. We are already seeing this trend in Chinese manufacturing companies. They are already adapting partial automation in many of their systems as the labor cost is going up there too.

One of the dominant characteristics of automation is replication. In the future, once humans achieve full automation in their manufacturing facilities, the location of these automated manufacturing units will become immaterial. It will be determined by the local market and not the availability of a work force. In the future, production units will be spread around the world. As a result, employees of every country will have to compete with the rest of the world for skilled automation jobs. The good news is jobs will be aplenty, putting more pressure on the education system to evolve.

Effect of White-Collar Job Outsourcing

Over the past few years, skilled labor has also been outsourced. Some research skills, which require PhD degrees, are so rare even in developed countries it is necessary to out-source them to other developing countries for faster research and development.

Other skills, including Information Technology, that do not need doctoral degrees, are outsourced for both skill and cost reasons. These jobs are moving from country to country depending on low cost and skill availability.

None of these skilled jobs has any national borders. In this digital age, the employee can be in any country and join the work force. With such global availability of labor and easy access via computer networks, it is relatively easy to set up a global organization. The headquarters can be in one country and its branches spread to several other countries. This kind of organizational structure brings in the best available skill at a low cost and helps grow the global market even larger.

We get a much better understanding of these white-collar jobs looking at the outsourcing from the evolutionary point of view. Consider the example of information technology (IT) outsourcing. It appears to threaten local job markets in developed countries. However, the reality is much different.

When IT jobs are outsourced, the company's capability increases. It can produce more and better software for the same investment. For instance, in place of one video game, the company now can produce three or four games for the same cost. This is because, the company can hire as many as four to five employees for the salary of one that is paid in a developed country like the United States or Europe. Instead of one version of software produced every two years, the out-sourcing company may produce a new version every three months. A GPS software in a cell phone may update maps every month instead of once every year or two.

More products fetch more revenue for the company, and as costs come down market size increases. To satisfy bigger markets, production has to go up making the company hire more employees overseas. More employees are also needed locally to manage the outsourced projects. This is how the positive feedback loop begins to take effect in all outsourced businesses.

Other competing companies outsource, too, and their capability increases. Because of increased market size, many new companies emerge and we see more employment opportunities both in overseas and local markets. A company producing ten software products a year with ten employees before outsourcing, could now produce 50 software products or more with the capabilities of outsourcing. As the market size increases, they now need more than ten employees locally and several hundred people overseas. This positive feedback loop keeps creating more jobs both locally and overseas. The number of jobs created locally will soon exceed the jobs lost. While this is just one side of the coin, there would be ripple effects on other industries as well.

With more jobs outsourced, there will be a tremendous need for infrastructure and networking in outsourced countries. With millions of people needing access via the network, the telecommunication and networking industries in developed countries will get a boost as they expand their capacity and hire local people to expand the telecommunication networks. In India for example all the 250,000 villages are now being connected by fiber optic cables. This is a massive undertaking. This not only advances the country digitally, it also helps globally as it becomes a mega market for cellphones and laptops. Who is benefitted by India going digital? Companies like Apple, Google, Facebook, all cellphone manufactures would greatly benefit from it as they have a huge market to sell their products. India gets benefited in a big way by increasing its standard of living. Globalization is always a win-win situation.

What is the impact on international shipping and travel? The ships and airlines will get their share as more goods are transported and more people travel all over the world for business meetings, sales/marketing, contract jobs, and so on. The spending from these visitors has also proportionately increased. According to the department of transportation, U.S. and foreign air carriers transported 142 million passengers between the United States and the rest of the world for the year-ended December 2000 and 190.5 million passengers for the year-ended December 2014. During the same time US GDP has grown from 10 trillion to 17 trillion and world GDP has grown from 50 trillion to 77 trillion. We are seeing more aircrafts being added to existing fleet in most airlines. We see job opportunities here too. What is the impact on international banking? With so many transactions going on in the world, the banks will need more people. Growth of just one industry will have a substantial effect on other dependent industries. There is a direct relationship between these developments to outsourcing.

If we look back and see how we evolved all these centuries, these kinds of developments take time. However, with outsourcing, the pace is so high that we are seeing all these effects in our lifetime.

Outsourcing or Mutual Sourcing

As underdeveloped countries develop, there will be more demand for education and training. The online education system that we discussed earlier will get a boost to support such a huge global demand and there will be availability of more skilled labor. In the future, as underdeveloped countries advance, it will be hard to say which country is outsourcing to which. The skill level being outsourced eventually will become *mutual outsourcing* spread all over the world.

When we look at the outsourcing crisis from evolutionary eyes, we see the future world as a network of intellectual humans commanding a host of machines that do all but the manual jobs. Blue-collar jobs will cease to exist in the future. White-collar jobs will be mutually outsourced from every country to the other. Outsourcing is just the beginning of that massive development.

CHAPTER 16

PLANNING YOUR CAREER

Planning a career is the hardest decision one can make. Choosing a job that does not have the potential to be outsourced or automated is as important as understanding the evolutionary phases of your job so you will not be laid off as the world progresses.

As discussed earlier, outsourcing happens to cut costs. In this era of globalization, no job is safe in either developed or developing countries. Even in fast-growing countries like China and India, the jobs have the potential to be re-outsourced to even lower-cost countries. The good news about outsourcing is it creates more local service jobs that are difficult to outsource.

To understand the job market, let us create three categories: jobs that are currently being outsourced, jobs that are hard to outsource, and jobs that have the potential to grow without being outsourced. [2]

---

2 Note: As this list changes from time to time, further research will be needed for the latest information.

Currently Outsourced Jobs

These examples include software programming, medical bill processing, medical transcription, accounting, call center and technical support, data entry, paralegal, tax preparation, and biomedical research.

This does not mean you cannot find jobs in these areas. It only means they have global competition, as they do not require an employee's physical presence to deliver the customer service. Any job that does not strictly require physical presence to interact with the customer has the potential to get outsourced. As IT improves, more jobs fall into this category.

White-Collar Jobs That Are hard to Outsource

This list includes physicians, lawyers, physician assistant, dental assistant, physical therapy, pharmacist, school/college teachers, fitness trainer, and sports coaches.

These professions need a physical presence to serve their customers and have very little potential to get outsourced. Again, this does not mean they will not be outsourced at all. Partial outsourcing is possible here, too. However, these hard to outsource jobs are potential targets for automation as businesses are constantly searching for ways to cut costs.

Healthcare, for example, is becoming very expensive. Health insurance companies will find ways to automate many processes to reduce cost. If you are planning entry into the healthcare profession, you need to evaluate your skill and ability to face the challenges when it is automated in the future.

We are already seeing the impact of automation in healthcare, driving doctors and other workers to learn additional technical skills every year to remain competitive. The number of diagnostic devices increases each year. Many new technologies for medical procedures are emerging onto the market. Documentation is becoming electronic. Accessing patient medical information via online has become fairly common in many developed countries.

Online monitoring of patients is now becoming the new trend to cut costs. Minimally invasive surgery is gaining popularity. If you look at this profession from an evolutionary angle, it has all the characteristics of being heavily automated in the near future.

As healthcare costs climb due to technology, cost reduction will become the focus for health insurance companies. Treating a sick patient is very expensive. Treating the same problem at an early stage is highly cost effective. Preventing the same ailment is much cheaper. The earlier symptoms are diagnosed, the more money the company saves. You will see a host of health monitoring devices in the future that collect and send patient data routinely. Availability of automated software and machines will increase to meet the demand.

In the same way, teaching jobs will face similar competition. Online colleges will increase in number in the near future. If you are planning to be in this profession, be prepared to meet the competition to become a successful online teacher or be prepared to be a research associate to support and maintain online teaching sites, which needs continuous learning to get the latest and greatest information to compete with other online institutions.

The bottom line is, in the future these professions, though they are hard to outsource, continually require hi-tech knowledge to stay competitive. For any of these professions, you will need an aptitude to adapt to ever-changing automation to be successful.

Blue-Collar Jobs That Are hard to Outsource
This last list includes plumber, electrician, auto repair, home construction, mail/courier, food service industry and so on.

Just like the previous category, these jobs also require physical presence to serve their customers and have very low potential to get outsourced. At the same time, they have the potential to grow and expand because of globalization. You may heave a sigh of relief at finding jobs that will expand in the future instead of collapsing like others. However, there is a catch. As these jobs cannot be outsourced, their demand increases when the globalization pace increases. As a result, skilled workers become harder to find, making them expensive. Plumbing, electrical, construction, and a host of other jobs become expensive as rapidly as globalization spreads. These jobs suddenly look very lucrative. Many people may even change careers to get into these high paying jobs. The question is whether they will remain lucrative?

If the cost of labor increases in any profession, automation will find an evolutionary solution for it.

Traditionally, when an appliance broke down, a service person living miles away came to your home only to find it needed a small fuse replacement. However, you were charged by the hour as they lost time, too. Now there are smart appliances that can be controlled remotely through phone apps. Many also have the capability of having a technician perform remote diagnostics, which lowers the overall cost.

Remote monitoring, remote maintenance of advanced home appliances will become a challenge for service personnel in the future. If you look at these safe jobs of today from an evolutionary angle, they are great candidates for automation. If you are planning your career in this service sector, plan on facing these automations.

The same logical explanation goes with automobile repair. Services become much more expensive as the technology of cars advances. Most cars will be diagnosed remotely to reduce costs. We already see remote monitoring technology in many cars, such as unlocking cars remotely. The oil change we do regularly may become very rare in the near future and fully obsolete in electric cars. Soon you will find maintenance-free cars that do not require any regular maintenance and the customers need not have to drive to the service station every three months. These cars will be remotely diagnosed for problems and remotely monitored, to check the car's health. If there is any potential problem, the system will know in advance and the customer will be contacted to have it fixed. Especially when self-driving cars hit the market, technician jobs will become technology intensive.

The Rule of Thumb

When it comes to picking the right career, look at it from both a globalization and automation point of view. The rule of thumb is that globalization will outsource jobs that do not require physical presence to serve the customer. Automation makes jobs that are hard to outsource technologically challenging, which means jobs will either get outsourced or will become technology intensive and move towards automation.

The bottom line is no matter what safe career you chose, it will eventually be impacted by automation. Irrespective of the profession you choose, a good understanding of Information Technology (IT) goes a long way in helping you to face the challenges. All automations are influenced by IT. Learning IT at the beginning of your career will help you understand automation better and keep you competitive throughout your professional life.

PART IV
FUTURE DILEMMAS

CHAPTER 17

BITTERSWEET FUTURE

Manual labor will cease to exist by the time we reach the end of the Age of Automation and humans achieve the real freedom of which they dreamed. Products will be manufactured, transported, and delivered to our doors by robots. Our food will be prepared by robots; homes built and managed by robots; and our health will be monitored by nanobots. There will be no hefty mortgages to pay, no grocery shopping to do—you will not even have to drive.

Life could not be any better.

However, it will all come with a hefty price. As the demand for quality will be sky high, life will be very hectic and stressful, making the need for highly technological products even more necessary, with less and less time to perform such motor functions ourselves.

There will be less and less human-to-human interaction in the future. We will expect more from machine-managed systems. As we deal with machine slaves the majority of the time, we will get used to expecting the best service all the time. If a product or service were ordered, the delivery time will be set not by days but by hours and minutes. Thanks to the invention of drones. More quality, more precision, timelier service—more of everything will mold our lives.

With the majority of jobs taken over by robots, the only jobs available to humans will be centered on machines and robots. Every profession including physicians, engineers, and lawyers will put their skills into computers, so machines will do their routine chores, allowing the professionals to concentrate on further growth. As competition becomes intense, job stress will be equally high, as will the rewards.

Robots will undergo continuous change and become more sophisticated as time goes by. The need for improved robots will never stop. Continuous growth is what technological evolution is all about. In the Age of Automation, robots will know so much and we will know so little we will not be able to make a living without them.

Who is responsible for such a fatal dependency? Every one of us who demands high quality and high precision will be responsible for such dependency on machines. Even in today's world, our dependency on computers, cell phones, televisions, and a host of other electronic gadgets is increasing every day. If we lose them, we feel their absence. The Age of Automation is no different. Machines will offer us a high quality life. It is inevitable we will follow this path.

What Motivates People to Work in an Affluent Future?

Our evolutionary future looks so rich and secure one wonders if we really need to work at all in the future and whether there will be motivation to do so. The question of why rich people work has been pondered ever since our ancient economy divided the rich and the poor.

It is understandable poor people struggle hard to make money. However, it is difficult to understand why the rich and famous, who have all the financial security in the world, work equally hard. But if you read about the lifestyles and activities of billionaires, they are even busier than most others.

Achieving more in one's lifetime is the secret behind our evolution. A millionaire tries to add one more million to it. A billionaire tries to double it. More money, more power, more fame keeps us busy. Everyone struggles hard to acquire more of everything.

In the rich future, however, people will have a choice. They can sit doing nothing or work in a high stress job. If they sit doing nothing and depend on social security, they will helplessly watch others progressing in their lives earning more money, recognition, and identity, and enjoying all the great joys of what the world has to offer. Eventually desire wins.

To increase the standard of living, the common person will work harder, and that will keep our economy going. Rich people will try to increase their wealth, which will boost the economy further.

Both rich and poor contribute to the growth of our economy. Our desire for more keeps us evolving technologically.

CHAPTER 18

GREEN TECHNOLOGY AND GLOBAL WARMING

Green technology has become very popular terminology out of sheer concern for our planet. We have welcomed this technology into our lives, even though there is very little or no economic gain.

Although we all have to embrace this technology, all green technologies are not necessarily fully green. We need to find out how much green the technology is even before considering using it. For instance, even though electric cars are greener than their gasoline counterparts are, they are not fully green as the company that manufactures the battery may emit greenhouse gases. The same argument goes with organic products. Organic crops use less or no pesticides and are very healthy. However, because they use little or no pesticides, the result is a low yield crop with high cost. To increase the yield, we need more land which results in more deforestation. Although organic crops are healthier and greener than their non-organic counterparts are, they are not fully green either. Does it mean companies are simply labeling some technologies as green even though they are not fully green? How green should a technology be for humans to consider it seriously green?

A true green technology is one that has no impact on our ecosystem, either at the production level or at the usage level. On the other hand, a greener technology is one that has lesser impact on our ecosystem compared to its non-green counterpart. We do not have true green technologies yet. Most of the green technologies we see around are only greener technologies. It is the green industry's responsibility to provide data proving how green they are. Now recently there is a usage of a gauge called *carbon foot print* that determines the green level. It is measured in terms of Tons of CO_2/year. In the United States, in the Midwest an average person living in a single family home is responsible for emitting about 68 tons of CO_2/yr. If proper green steps are taken to use efficient heating and lighting, use energy star appliances then the carbon foot print will reduce to reduce to 40 tons/yr. People can also voluntarily contribute money to nature conservatory to help offset their carbon foot print. Your voluntary contribution to The Nature Conservancy's carbon offset program will help fund forest conservation, reforestation projects that produce measurable climate benefits.

To measure greener technologies, there are also many other gauges like toxic waste produced, recycling methods used, disturbance to natural habitat etc.

Now let us look at green technology from an evolutionary angle. There is a lot more meaning to green technology than what the definitions says.

Life on this planet is so dependent on the ecosystem that without it we cannot survive. Our planet cleans the air we breathe, the water we use, digests all the garbage we discard, and feeds us with flora and fauna on which we depend. Our planet feeds us and cleans us like a mother looks after her baby. This mother-child relation is not easy to break. It is the umbilical card connecting humans to Mother Nature. To break this cord and go free like any child that grows and matures, we need to develop a highly sophisticated technology that can mimic earth's ecosystem. This technology is nothing but 'green technology' that we are all talking about and striving to achieve. Any technology that can produce products without affecting Mother Nature is a true green technology.

All human space voyages, including lunar landings, are examples of man's quest to explore the universe by breaking the dependency on earth's ecosystem, however short the duration may be. It is like a baby walking away from its mother and coming back after a few little baby steps…a few more steps, and then some more.

We are in that stage now, craving independence and not afraid to stray too far from Mother Nature, but cautious enough that we are not quite ready to break ties completely. Every rocket that carried humans to space tried to mimic our ecosystem for a short period of time. That same technology will grow and improve over time, making us better able to move away from Mother Earth.

When we are ready to break that cord and start living an eco-free life either on this planet or elsewhere, which will eventually happen when technology matures, we should be able to create our own artificial atmosphere. Our industries have to be fully emission free. In the future, we will be utilizing all the green technology we are developing today to keep this artificial atmosphere clean.

Our generation is already talking about colonizing Mars. Our next generation may even achieve that, but not without green technology. In fact, green technology will be so vital in these artificial atmospheres that without it we can't possibly survive. For instance, if there is carbon dioxide emission from a car, our eco-system absorbs it. In an artificial atmosphere like a Mars colony, it will be catastrophic. The green technology we are developing today is one of the greatest contributions we can give to our future generations to cut that umbilical card with Mother Nature and live independently.

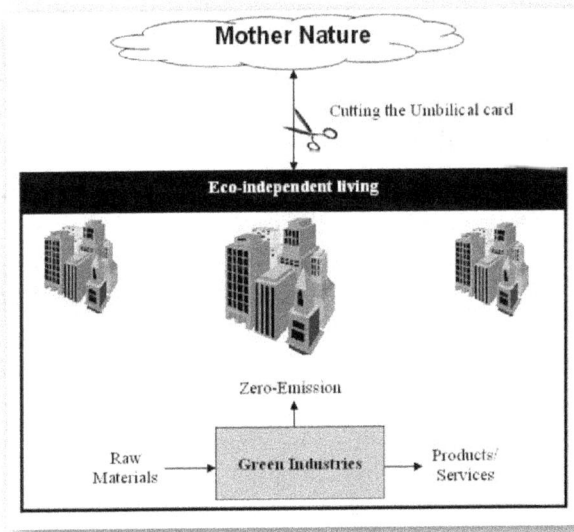

Fig. 16: Eco-independent living in the future

If our current global warming phenomenon had not been observed and had we not developed green technology, we probably would have developed it regardless when we tried to break away from Mother Earth. Now, thanks to global warming, we are forced to develop this technology much earlier. Green technology will help us to evolve from an eco-friendly world to an eco-free world.

Automation and Global Warming

Let's look at other extinction level threats that humans face. Any celestial disturbance, close to home, is an extinction threat to humankind. Meteor strikes, gamma ray bursts, black holes—the list of catastrophes that could wipe us out is endless. However, the good news is all of these forces are highly predictable, following the laws of science. What we need is the right technology to do so.

While we are concerned about science's endless theories of how the world will end, with continuous technological growth, one day we will be able to predict and manipulate potential celestial threats. However, to get to these threats, before they get to us, it is a race against time. Although more time allows us to be more knowledgeable and stronger against these threats, we do not have unlimited time. Every economic slowdown works against us. *Every economic recession takes us closer to extinction.*

Global economy is the only shield that will protect us in the future. Protecting our economy is the same as protecting the human race. While we always look at the economy as a way of making quick money and spending it in our short lives, in reality, our future survival is in the hands of global economy.

Many argue research and technology will make us stronger. The truth is, our ability to use technology is as important as technology itself. For instance, if our researchers develop a technology to launch a gigantic space mission to divert an oncoming meteor on collision course with our planet, will we have the financial capability to manage such a huge mission? Only an equally huge global economy could handle it. Similarly if our engineers develop a plan to rebuild a city that got destroyed by a hurricane, will we have the financial capability to manage such a huge project? Only an equally huge economy could handle it. If our economy is weak, so are we.

Though the Age of Automation will give us the ultimate power to survive these threats, how quickly we reach it determines our survival possibilities. Although the Age of Automation appears to be far off, the pace to reach it is in our own hands. If we understand the evolutionary nature of our global economy, and how fast we could grow it, the Age of Automation appears closer.

With a robust economy, nations can help each other during bad times. If a particular country were badly hit by a disaster, only an economically developed country could help them out. If all nations are weak, neither can they help each other nor can they help themselves. This gives us even more reason to continue building a robust global economy and not just individual economies.

How Seriously Should We Consider Global Warming?

Global warming is the century's most debated subject, mainly because of our concern for the planet and our own existence. Many strongly argue global warming is not really happening, and there is not enough evidence to prove it. The question is, do we have time to analyze all the current and prehistoric data and reach a conclusion? Do we have to wait for our ecosystem to show its ugly face and start wiping us out? If scientists do prove global warming is a real threat, how much time do we really have, to develop technologies to save ourselves? We have to grow our global economy and its technological capability at a higher pace to survive. The question now becomes: At what pace should we grow? What pace is right for us? Can we afford the luxury of a slow growth that we are currently going through? Global economic development and technological evolution are very natural processes. However, we need to regulate and monitor to eliminate all economic bubbles and the energy bottleneck to avoid economic recessions and achieve significant technological advancement.

What Can We Expect From Global Warming?

In spite of the fact that a major chunk of our atmosphere is brutally cold, we are still able to lead a comfortable life on the ground because of a stable atmosphere. Any disturbance to this steady atmosphere will expose us to more cold. More freezing temperatures is what we can expect as this disturbance intensifies. Normally we imagine a warm climate when we think of global warming, but in reality it is cold that we should be concerned about.

Global warming is all about a global weather disturbance due to temperature rise. The actual temperature rise is only a degree or so as claimed by many scientists around the world. So what is the big deal? Here are the effects of global temperature rise. Firstly the water evaporation rate from our oceans increase because of this temperature rise even though it is just one degree. As you may know two thirds of earth's surface is water. So the surface area of evaporation is enormous. It is about 360 million square kilometers. A degree rise in temperature will produce lot of moisture that form clouds and it has to fall eventually. When it falls, we see more rain or more snow. The second factor is that the rise in temperature will alter the course of seasonal wind direction. Even if the deflection is small, if you look at the surface area of earth, the over long distances, it will be significant. Which means, some areas may not get any rain at all, or some areas that never received rain before will start getting more rain or snow. We are already seeing these effects all over the world. The third and most important factor is the melting of ice in arctic zone, which will rise sea level affecting all coastal cities.

As the clouds increase because of global warming, they begin to cover more of the earth's surface and we may get some relief due to global cooling. How much will this cooling offsets the global warming is yet to be seen. May be at some stage the atmosphere will balance again and become steady again. Our understanding of atmosphere is so primitive, we don't know where that balance point is and when it will happen. Until then we have to endure the fury of the nature.

Severity Levels of Global Warming Threat

The severity of global warming depends on the level of technological evolution at the time the effects present themselves. For instance, in the current age even a category two or three hurricane appears very powerful. Hurricanes and tornadoes regularly kill humans and destroy property nearly every time they hit. It appears we are at the mercy of these natural forces.

Ordinary effects of global warming appear very powerful in the current age. Many parts of the world will experience either draught or flood. Coastal cities may go under water and may need evacuation. We will be hit very hard economically. We should be prepared to deal with it. This is where our automation would come to our rescue.

In the Age of Automation, however, we will not feel as threatened by these effects. The monstrous machines will help us survive from the hostile environment and we will feel much safer. We will also be skillful enough to build our own artificial world inside this natural world. We will not only survive, but also continue to evolve, and global weather will become inconsequential.

A Race with Global Warming

If global warming effects start threatening us in the current age, our economy will slow down and we will go into survival mode. Survival mode will not actually keep us alive for very long, because it will put an end to our economic growth. With a very weak global economy and severe global warming effects working against it, our survival will be at stake. Rather than playing defense and waiting for these events to happen, we should try our best to outpace the speed of global warming. If our progress into the Age of Automation occurs more rapidly, possibly even well before global warming becomes severe, we will have a higher probability of survival and continued evolution.

No one is sure if global warming could be reversed. By using green technology, we may reduce the pace of global warming to some extent. However, pace of our technological growth could be enhanced significantly.

If the global economy and global warming race side by side, a likely scenario considering the speed at which our economy is currently growing, we will be cutting it very close and taking chances with our very survival.

CHAPTER 19

WHAT WE ALL CAN DO

Automation needs a very highly skilled labor force that our current education system is not producing quickly enough. Our world needs help from every one of us to produce the needed skilled force at a faster pace.

We can all contribute to global growth. If you are in school, plan your career after considering the effects of both globalization and automation. Every job faces global competition in the future. Remember, if it cannot be outsourced, it eventually will be automated, which happens at a much higher pace in developed countries than in underdeveloped ones. No matter which part of the world you live in, you need to consider both aspects for career planning.

If you are planning to drop out of school, think again. With global competition, you will not go far. If you are dropping out because of financial reasons—a likely scenario for many—consider online training. Online institutions are the future of education system and you will see a surge of activity in the near future.

If you are in college, make sure you learn cutting-edge technology to begin your career. Choose your electives wisely. They have to be productive and relevant.

If you are already in the work force, anticipating change and continuous training are key elements to keeping your job. No matter how current your skills are and what cutting edge technology you have acquired, both change over time. It is not uncommon to see people laid off as the economy fluctuates. If you are keeping up your skills, finding another job will not be a problem. In the future, this pressure will be much greater.

Any awareness of future automation will not only make you proactive at work, but also an asset to the company. Be ready for changes in your workplace, as competition between companies grows more intense in the future. You will be much better off if you can anticipate those changes yourself rather than being surprised when they happen.

Changes in the workplace used to happen on a very gradual pace. Now, because of the surge in globalization, changes are happening rapidly and we need to understand their necessity to plan ahead. Well-informed employees are an asset to organizations when such changes happen. We have seen in many cases employees resisting change; they want to do things the way they have always been done. This puts a restriction on the organization's growth to stay competitive. Both the organization and the employees should understand the evolutionary path the world is treading.

If you are a parent, knowledge of the technological evolution helps you guide your children in the right direction. Let your children know how your own job is changing, what kind of global competition you are facing in your own career, and what have you planned to face it. This will help your children understand the technological evolution you are going through and what they can anticipate in their careers.

If you are retired, try applying the five strategies discussed in this book to the career you have just retired from and see how they changed over time and how the jobs will be different in the future. You will become a well-informed guide to your children and grandchildren in planning their future.

If you are a philanthropist, you can help the world by promoting online education centers in underdeveloped countries. Remember, the more children you enroll into the education system the more robust the global economy will become. While you enjoy the ethical satisfaction of educating the world's children, the world enjoys the little push you gave for global growth.

CONCLUSION

AN ART WE ALREADY HAVE

Not all countries are developing at the same pace. We still have poverty, hunger, and illiteracy in the world. We must realize all the present day problems will gradually fade away as our global standard of living improves.

Once poor regions of the world prosper from globalization and start leading comfortable lives, their focus will change from violent living to a safer and longer life. If we progress slowly because of economic recessions and lack of skilled people, we will continue to face internal conflicts. Until the world reaches the Age of Automation, the pace of technological evolution is paramount.

Though you may not agree to all the ideas and concepts, the goal of this book was to make you look at world problems from an evolutionary angle. The knowledge of technological evolution helps us visualize the evolutionary forces responsible for a fast-paced, global economic growth. The concepts of globalization and automation help us look at our own life changes from a different perspective.

After all, the art of looking into the future is an art we already have; we only need to master it.

REFERENCES

Centers for Medicare and Medicaid Services, "National Health Expenditure Data: NHE Fact Sheet" https://www.cms.gov/NationalHealthExpendData/25_NHE_Fact_Sheet.asp

Congressional Budget Office - Healthcare expenditure: http://www.cbo.gov/ftpdocs/89xx/doc8947/MainText.3.1.shtml

Centers for Disease Control and Prevention. Chronic Disease Overview.

Dow Jones Reprint Solutions. http://www.djreprints.com/inkonpaper/iop-artimages.html

Macroeconomic Assumptions: USDA - United States Department of Agriculture http://www.ers.usda.gov/Briefing/baseline/macro.htm

Congress of the United States, Congressional Budget Office (January, 2008). Technological Change and the Growth of Health Care Spending. http://www.cbo.gov/ftpdocs/89xx/doc8947/MainText.3.1.shtml

Joint Economic Committee United States Congress, retrieved from http://www.house.gov/jec/news/Housing%20Bubble%20study.pdf

Financial Crisis Inquiry Commission THE MORTGAGE
CRISIS -APRIL 7, 2010 http://www.fcic.gov/reports/
pdfs/2010-0407-PSR_-_The_Mortgage_Crisis.pdf

U.S. Energy Information Administration, Department of
Energy, retrieved from http://www.eia.doe.gov/emeu/
aer/pdf/pages/sec11_2.pdf   http://www.eia.doe.gov/
emeu/aer/pdf/perspectives_2009.pdf

U.S. Energy Information Administration/Annual Energy
Review 2009. Energy bubble before the recession.
http://www.eia.doe.gov/emeu/aer/pdf/pages/sec11_
14.pdf

NASA, "Water Vapor Confirmed as Major Player in Climate
Change," retrieved from http://www.nasa.gov/topics/
earth/features/vapor_warming.html

International Monetary Fund - World Economic Outlook-
Sustaining the Recovery October 2009 - http://www.imf.
org/external/pubs/ft/weo/2009/02/pdf/c1.pdf

Global Health workforce by density,-World Health Organization
http://www.who.int/whr/2006/overview_fig6_
en.pdf

U.S. Bureau of Labor Statistics, U.S. Department of Labor.
100 years of consumer spending. http://www.bls.gov/
opub/uscs/report991.pdf

The National Center for Public Policy and Higher Education. Measuring up 2006: The national report card on higher education. http://www.highereducation.org/reports/mup_06/MUP-06.pdf

Structure of earth. Retrieved from http://scign.jpl.nasa.gov/learn/plate1.htm

Earth's crust. Retrieved from http://earthquake.usgs.gov/research/structure/crust/index.php

"Technological Change and the Growth of Health Care Spending." A CBO Paper - January 2008. U.S. Congress – Congressional Budget Office based on data from the National Center for Health Statistics. Fig. 10.

World Health Organization. Global Atlas of the Health Workforce. (http://www.who.int/globalatlas/default.asp). Fig. 11.

U.S. Energy Information Administration, Annual Energy Review 2009: Energy Perspectives: 1949-2009 Report No. DOE/EIA-0384(2009): Release Date: August 19, 2010. Fig. 12.

The U.S. Housing Bubble and the Global Financial Crisis: Housing and Housing-Related Finance. -By Ranking Republican Member Jim Saxton (R-NJ) - Joint Economic Committee - United States Congress May 2008. Fig. 6.

International Monetary Fund, World Economic Outlook-Sustaining the Recovery October 2009 (Chapter1 Global Prospects and Policies - P5, Figure 1.3. Developments in Mature Credit Markets) Permission to publish: Ref#3716498 date: 12-15-10. Fig. 7.

International Monetary Fund, World Economic Outlook-Sustaining the Recovery October 2009- (Chapter1 Global Prospects and Policies – P50, Appendix 1.1, Figure 1.16) Authors: Kevin Cheng, Nese Erbil, Thomas Helbling, Shaun Roache, and Marina Rousset. Permission to publish: Ref#3716498 date: 12-15-10. Fig 8.

Federal Housing Finance Agency: January 1980-100. Fig 13.

U.S. Department of Labor, Bureau of Labor Statistics. Fig 15a.

Congressional Budget Office, based on data from Bureau of Labor Statistics. Fig 15b.